A TEXT BOOK

OF

NATURAL PHILOSOPHY:

AN

ACCURATE, MODERN, AND SYSTEMATIC

EXPLANATION OF THE

ELEMENTARY PRINCIPLES OF THE SCIENCE.

ADAPTED TO USE IN

HIGH SCHOOLS AND ACADEMIES.

WITH 149 ILLUSTRATIONS.

BY LE ROY C. COOLEY, A. M.,

PROFESSOR OF NATURAL SCIENCE IN THE NEW YORK STATE NORMAL SCHOOL.

"Let science, by cultivating man's intellect, elevate him to nobler and more spiritual views of God's wisdom and power."—*Cooke.*

NEW YORK:

CHARLES SCRIBNER & CO.,

654 BROADWAY.

1869.

ALVORD, PRINTER.

TOPICS FOR REVIEW.

PART I.

THE PHENOMENA OF MATTER AT REST.

INTRODUCTION.

The Properties of Matter.—(1.) Properties of matter—Extension—Impenetrability—Indestructibility—Elasticity. (2.) Physical Properties—Chemical Properties. (3.) Natural Philosophy.

CHAPTER I.

§ 1. *The Fundamental Ideas.*—(4.) Molecule—Inertia—Attraction—Repulsion. These four ideas explain the phenomena of nature.

§ 2. *Varieties of Attraction.*—(5.) I. Gravitation — Universal — The first law—The second law. II. Cohesion—Its power—Through insensible distances. III. Adhesion. IV. Capillary Force — Causes liquids to penetrate porous solids—The first law—The second law.—(6.) These varieties of attraction are but different manifestations of a single influence—Problems.

CHAPTER II.

OF THE PHYSICAL FORMS OF MATTER.

Introduction.—(7.) Application of the fundamental ideas.

§ 1. *Of the Characteristic Properties of Solids.*—(8.) Hardness—Tenacity—Malleability—Ductility—Crystalline Form.

§ 2. *Of the Characteristic Properties of Liquids.*—(9.) Attraction and repulsion nearly equal—Mobility.

§ 3. *Of Liquids at Rest.* (10.) Pressure equal in all directions—Surface is level—The level surface is convex. (11.) Water in pipes rises as high as its source—The supply of water to cities—Springs—Artesian wells. (12.) Pressure independent of the shape of the vessel—It depends on the depth of the liquid—To calculate the pressure.—(13.) Bodies immersed are pressed upward—With force equal to the weight of fluid displaced—The solid lighter than water—The solid heavier than water. (14.) Specific gravity—I. Of gases—II. Of liquids—By direct weighing—By the hydrometer—By the use of a bulb—III. Of solids—Heavier than water—Lighter than water—Table. (15.) The equal transmission of pressure—The shape of the vessel makes no difference—The hydrostatic press—Problems.

§ 4. *Of the Properties of Gases.*—(16.) Compressibility—Expansibility—Elasticity—Weight.

§ 5. *Of the Pressure of the Atmosphere.*—(17.) The atmosphere exerts pressure in all directions about 15 lbs. to the square inch. (18.) I. The barometer shows the pressure of the atmosphere—The pressure of atmosphere depends upon its weight—Upon the amount of water-vapor it contains—Upon the elasticity of its lower portions. II. The Common Pump. III. The Forcing Pump. IV. The Siphon.

§ 6. *Of the Relation between Volume and Weight.*—(19.) The volume of a given weight of air depends upon pressure and temperature—I. Pressure—Volume inversely as the pressure—Density of the atmosphere—Is greatest at the surface of the earth. II. Temperature—Heat increases the volume of air $\frac{1}{490}$ of its bulk for each degree—Problems.

PART II.

THE PHENOMENA OF MATTER IN MOTION.

CHAPTER III.

OF MOTION.

Introduction.—(20.) Application of the fundamental ideas.

§ 1. *Of Motion caused by a Single Force.*—(21.) The first law of motion—The second law—The third law. (22.) Velocity—Uniform velocity—

Impulsive force—Uniform velocity produced by an impulsive force—Space equals time multiplied by velocity. (23.) A constant force—Uniformly accelerated velocity—Difficulties in the way of experiment—Overcome by Atwood's machine. (24.) Experiments with Atwood's machine—Proof of first principle—Proof of second principle—Analysis of the motion of a falling body—Construction of the table—From the table obtain the laws—From the table obtain the formulas—By the formulas solve problems—Problems.

§ 2. *Motion caused by more than one Force.*—(25.) If a body be acted on by two forces represented by the sides of a parallelogram, they are equivalent to a single force—The resultant may be found.—(26.) Any force may be resolved—To find the component which acts in a given direction. (27.) Two forces in the same direction—The point of application—The weight of a body—The center of gravity. (28.) Curved motion—Is produced by at least two forces—One of which is constant—Projectiles—Their motion due to two forces.

§ 3. *The Indestructibility of Force.*—(29.) Force is indestructible—Motion can not cease without exerting the same amount of force which produced it—Suppose a body move without resistance, *Momentum*—Motion due to an impulse meets with resistance—Suppose a constant force applied to overcome resistance, *Living Force.*

§ 4. *Of Machinery.*—(30.) The principle of momentum—Machines—Simple—The law of equilibrium. (31.) Levers—Three classes—Application of the principle of momentum—the compound lever—Application of the lever. (32.) The wheel and axle—Acts on the principle of the lever—Application of the principle of momentum—One wheel may turn another by means of cogs—By friction—By bands—Application of the wheel and axle. (33.) The pulley—Fixed and movable—The principle of momentum to the fixed pulley—To the movable pulley with one rope—To the movable pulleys with separate ropes—Applications of the pulley. (34.) The inclined plane—If the power act parallel to the length of the plane—If it act parallel to the base of the plane—Applications of the inclined plane. (35.) The wedge. (36.) The screw—Application of the principle of momentum—Applications of the screw—Problems.

§ 5. *Of the Motion of Liquids.*—(37.) The velocity of a jet of water the same as that of a falling body—The velocity depends upon the depth of the orifice—Calculated by the formula $V = 2\sqrt{Sg}$. (38.) To calculate the quantity. (39.) The velocity less than the theory gives—The quantity less than the theory gives—Quantity increased by using tubes.

(40.) The undershot wheel—The overshot wheel—The breast wheel—The American turbine.

§ 6. *Of the Motion of Air.*—(41.) Wind—The trade winds—Due to heat and the rotation of the earth.

CHAPTER IV.

OF MOTION.—VIBRATIONS.

Introduction.—(42.) Application of the fundamental ideas.

§ 1. *Of Vibrations of the Pendulum.*—(43.) The pendulum—Vibrates under the influence of gravitation and inertia—The first law—The second law—The third law. (44.) The center of oscillation—The laws apply to this point. (45.) The pendulum is used to measure time—To determine the form of the earth.

§ 2. *Of Vibrations of Cords.*—(46.) The vibration of cords—Due to elasticity and inertia—The laws of vibration—first, second, third.—(47.) Progressive vibrations—The motion appears to be lengthwise of the tube—Ventral segments. (48.) The cord vibrates as a whole—In ventral segments at the same time.

§ 3. *Of Vibrations of Liquids and Gases.*—(49.) Water waves—May interfere. (50.) The vibrations of air—Alternate rarefactions and condensations—In all directions—Different sets interfere.

§ 4. *Of Vibrations of Molecules.*—(51.) Molecules in motion—Molecular vibrations affect our senses.

CHAPTER V.

OF THE EFFECTS OF VIBRATIONS.—I. SOUND.

§ 1. *The Origin and Transmission of Sound.*—(52.) Sound produced by vibrations. (53.) Sound waves—Are transmitted through all elastic bodies—With different velocity in different media—The first law—The second law—With uniform velocity in the same medium.

§ 2. *Of Refraction and Reflection of Sound.*—(54.) Sound waves pass from one medium to another—Refraction—Sound made louder by refraction. (55.) The reflection of sound—The law—The echo.

§ 3. *Of Musical Sounds.*—(56.) Musical Sounds—Any noise repeated rapidly causes a continuous sound—Puffs of air made rapidly give a musical sound. (57.) Musical sounds differ—I. Pitch—Pitch depends on rapidity of vibration—Intervals—The diatonic scale—The number of vibrations for the notes. II. Intensity. III. Quality.—(58.) Musical instruments—Stringed instruments—Pitch raised by using strings of different lengths—Of different tension—Of different weight.

CHAPTER VI.

OF THE EFFECTS OF VIBRATIONS.—II. LIGHT.

§ 1. *Of the Nature of Light, and the Laws of its Transmission.* (59.) Light is the effect of vibrations—The ether—Luminous bodies. (60.) Rays of light—Are transmitted—In straight lines—With uniform velocity —Its intensity inversely as the square of the distance—Photometry.

§ 2. *Of the Reflection of Light.*—(61.) Reflection—The law. (62.) Mirrors—The effect of plane mirrors—The effect of concave mirrors—The effect of convex mirrors. (63.) Images by reflection—Image of a point. (64.) Images by plane mirrors. (65.) Images by concave mirrors — The image of a point. (66.) Object beyond the center — Between the center and the focus—Between the focus and the mirror. (67.) Images by convex mirrors.

§ 3. *Of the Refraction of Light.*—(68). Refraction—The first law—The second law. (69.) The index of refraction. (70.) Lenses—The effect of convex lenses—The effect of concave lenses. (71.) Images are formed by convex lenses—The image of a point. (72.) Object at twice the focal distance—Farther away—At a less distance—Between the focus and the lens. (73.) Images by concave lenses.

§ 4. *Of the Decomposition of Light.*—(74.) Prisms—Prisms refract light—Prisms decompose light. (75.) The black lines of the solar spectrum—The bright lines in spectra of artificial light. (76.) Decomposition by rain-drops—The primary rainbow—The red band on the outside—The colors in the form of an arch—The secondary law. (77.) Decomposition by refraction—The color of bodies—The color of the sky—The color of the clouds.

§ 5. *Of Optical Instruments.*—(78.) The microscope—The telescope—The magic lantern—The camera obscura—The eye.

CHAPTER VII.

THE EFFECTS OF VIBRATIONS.—III. HEAT.

§ 1. *Of the Sources and Nature of Heat.*—(79.) The sources of heat—

I. The heavenly bodies—II. Mechanical action, friction—III. Chemical action. (80.) The material theory—The dynamic theory.

§ 2. *Of the Transmission of Heat.*—(81.) Rays of heat—Transmission of heat rays—Laws of transmission—Law of reflection—Law of refraction. (82.) The diffusion of heat—I. By conduction—Through solids—Through liquids—Through gases. II. By convection—Air is heated in no other way—Liquids are heated by convection—No convection in solids. III. By radiation—It depends upon temperature—On the nature and condition of the surface.

§ 3. *Of the Effects of Heat.*—(83.) The action of heat is twofold—It raises temperature — It expands bodies — Solids — Liquids — Gases—Temperature is measured by expansion—The thermometer—Various forms. (84.) Temperature indicates the rapidity of molecular motion—Expansion indicates a change in the relative position of molecules—Sensible and latent heat—Specific heat. (85.) At the melting point—Temperature stops rising—But the expansion increases. (86.) The boiling point—Depends on the purity of the liquid—On the nature of the vessel—On pressure—At this point temperature is constant—But the expansion increases. (87.) Heat is required to expand a body—The same amount is given off when the body again contracts.

§ 4. *Of the Steam-Engine.*—(88.) The elastic force of steam—The boiler—The cylinder—The crank—High and low-pressure engines.

CHAPTER VIII.

OF ELECTRICITY.

Introduction.—(89.) Application of the fundamental ideas.

§ 1. *Of Frictional Electricity.*—(90.) Electricity produced by friction—The electrical machine—Electricity detected by electroscopes—Two opposite forces—Measured by electrometers—The first law—The second law. (91.) A charged body—A non-conductor—An insulated body—A charged body polarizes an insulated conductor. (92.) A series of conductors polarized—The theory of induction. (93.) The Leyden-jar—It may be charged—It may be discharged—The Leyden battery. (94.) The electricity of the atmosphere—Of the same nature as frictional electricity—Lightning is the discharge of oppositely charged clouds—The aurora—Produced by electric discharges. (95.) The effect of points—Lightning-rods. (96.) The mechanical effects of electricity—Chemical effects—Physiological effects.

§ 2. *Of Magnetic Electricity.*—(97.) Magnets—Natural—Artificial—Different forms—Their force strongest at the ends. (98.) Attraction and repulsion—The law. (99.) A magnet will polarize a bar of iron—Several bars in succession—All the molecules of a magnet are polarized. (100.) A bar magnet supported—Will point north and south—Its variation—Annual and daily. (101). The dip of the needle.

§ 3. *Of Voltaic Electricity.*—(102.) Voltaic electricity—The voltaic circuit—Grove's battery—Bunsen's battery. (103.) It is produced by chemical action—Resistance—Quantity and intensity—Quantity depends on the size of the plates—Intensity depends on the number of plates.—(104.) The effects of electricity—Heat—Light—*Magnetism*, the electric telegraph—Induction, the Ruhmkorf coil. (105.) Conclusion.

TO THE PUPIL.

MY DEAR FRIEND:—

THE most faithful efforts are sometimes followed by partial success, or, it may be, utter failure, for want of some proper method of study. I have tried to systematize the principles of Natural Philosophy, and I believe that if you reduce your efforts to a corresponding system, you will find the acquisition of this science less difficult, more pleasant, and of enduring value.

Let me suggest the following plan. First, read the heading of the paragraph (numbered in parenthesis) that you may know what subject the paragraph presents. Then, look over the topics (numbered without parenthesis) and compare them with the heading, to see what are its essential thoughts. After this, study each topic in order; and finally, *learn* the heading and see that you can develop its topics without referring to the book.

Do not be satisfied with the statement of facts alone; but carefully study the relation of thoughts. You will find the analytical contents valuable for this purpose.

Finally, I hope you will not regard this little book as a commentary on the subjects it treats. You will find it profitable to have some larger work at hand in which you can find additional explanations of subjects which are of special interest to you. I have not aimed to exhaust the subject, but to give you an outline which, by present study or future reading, you will be able to fill.

That your pleasure in this study shall equal that which I have felt while instructing so many classes in this science, is the desire of your friend.

L. C. C.

PREFACE.

This volume is designed to be a *text-book* of natural philosophy suited to the wants of high schools and academies.

The author believes that the following features of his work adapt it to the purpose for which it is designed.

1. It contains no more than can be *mastered* by average classes in the time usually given to this science. To this end, the polarization of light, sounding flames, and kindred subjects of a less elementary nature, are omitted. But that the pupil may have access to such important and interesting matter, an appendix has been added.

2. It presents a judicious selection of subjects. Omitting what ever is merely novel or amusing, it gives a plain and concise discussion of *elementary* principles, of theoretical and practical value.

3. It is an expression of modern theories. It recognizes the fact that the spirit of a new philosophy pervades every department of science, and presents the doctrines of molecules and of molecular motions, instead of the old theory of imponderables, which has been swept away. Carefully avoiding whatever is yet only probable, it seizes upon what has come to be universally accepted, and, as far as may be, adapts it to the course of elementary instruction which it proposes.

4. It is logical in the arrangement and development of subjects. A single chain of thought (see Analytical Contents) binds the different branches of the science into one system of related principles.

5. It is thoroughly systematized. Chapters, sections, para-

graphs, and topics, have been arranged with careful regard, on the one hand, to the relation of principles to each other, and on the other hand, to the best methods of conducting the exercises of the class-room.

At the beginning of each paragraph is a plain and concise statement of useful facts and principles, while the paragraph itself contains the discussion of them by topics in their natural order.

The mind can not work intelligently unless it has some object toward which to direct its efforts. No scientist pursues his researches by experiment, without first proposing some fact, or principle, to be tested. The discoveries of the immortal Faraday were drawn from experiments, not made at random, but conceived and executed to test the truth of theories proposed in his own mind *beforehand*. (See Faraday as a Discoverer, by Tyndall). The synopsis, at the beginning of each paragraph in this volume, gives the pupil a clear idea of the work proposed to be done. He is then prepared to see how the facts of observation may be used to establish the principles of physical science.

Moreover, there is an increasing number of teachers who believe that oral instruction is quite as important to the pupil as the study of a text-book. These headings of the paragraphs are *texts*, which, taken together, give a compact view of the entire science, and which will enable the teacher to freely supplement the discussions of the book, by experimental or mathematical proofs. To facilitate this work still further, references have been given to the most accessible and reliable works wherein the subjects of the text are more exhaustively treated. The works chiefly referred to, are: Silliman's Physics, Cooke's Chemical Physics, Atkinson's Ganot's Physics, Tyndall's Lectures on Sound, and Tyndall's Heat as a Mode of Motion. No teacher of Natural Philosophy can afford to be without these books.

ANALYTICAL CONTENTS.

PART I.—THE PHENOMENA OF REST.

(1.) The qualities of matter are called its properties.

(2.) All properties of matter are either *Physical* or *Chemical.*

(3.) Natural Philosophy is the science which treats of the *physical* properties of matter.

(4.) The FUNDAMENTAL IDEAS in Natural Philosophy are expressed by the words *molecule, inertia, attraction* and *repulsion.*

Attraction is called gravitation, cohesion, adhesion, and capillary force, according to the circumstances under which it acts.

(7.) *Attraction* and *repulsion,* acting upon the *molecules* of bodies, produce the three physical forms of matter: *solid, liquid,* and *gaseous.*

The characteristic properties of solid bodies are hardness, tenacity, malleability, ductility, and crystalline form.

The characteristic property of liquid bodies is mobility.

The characteristic properties of gaseous bodies are expansibility and compressibility.

PART II.—THE PHENOMENA OF MOTION.

(20.) Read (4) and (7). *Attraction* and *repulsion,* acting upon *masses* of matter, determine their condition of rest or *motion.*

Motion is uniform if produced by an impulsive force.

Motion is uniformly accelerated if produced by a constant force.

Motion is curved if produced by two forces, one, at least, of which is a constant force.

The *force* which causes motion will be reproduced when the motion stops: hence the principle of *momentum*.

The principle of momentum, applied to any one of the *simple machines*, will determine its law of equilibrium.

The free motion of *liquid* bodies is due to the attraction of gravitation, and must obey the laws of this force.

The free motion of *air*, or wind, is due to the action of heat.

(42.) Read (4), (7), and (20). *Attraction*, *repulsion*, and *inertia*, acting upon *masses*, or upon *molecules*, produce *vibration*—

Of the pendulum.
Of cords.
Of liquids.
Of gases.
Of the molecules of all bodies.

These vibrations of molecules affect our organs of sense, and give rise to the phenomena—

Of sound.
Of light.
Of heat.

(89.) A constant and opposite action of *attraction* and *repulsion* among the *molecules* of bodies, gives rise to the phenomena of electricity.

PART I.

THE PHENOMENA OF MATTER AT REST.

NATURAL PHILOSOPHY.

INTRODUCTION.

THE PROPERTIES OF MATTER.

(1.) The qualities of matter are usually called its properties. Those most important for us to notice in the outset are Extension, Impenetrability, Indestructibility and Elasticity.

1. *The Properties of Matter.*—In what respects is a block of granite so unlike a block of wood? The granite is brittle, it may be chipped with a chisel: the wood is soft, it may be cut with a knife. The granite is heavy; to lift it may require the power of an engine: the wood is much lighter; perhaps a single arm may move it. We are thus able to perceive a difference in bodies only because there is a difference in the qualities they possess. These qualities are called *properties.*

2. *Extension.*—Every body of matter, however small, fills a portion of space. It is not possible to think of a body which should have no size. This property of matter, by virtue of which it occupies space, is called *extension.*

3. *Impenetrability.*—Not only do all bodies occupy space, every body fills the space assigned it to the exclusion of all others. One body may not be pushed into the substance of another; it can take the place of another only when the other has been thrust away. When, for example, a nail is driven into wood, it pushes the particles of wood out of its way; and when the hand is plunged into water, the water is thrust aside to give it place. This property of matter, by virtue of which no two bodies can fill the same space at the same time, is called *impenetrability.*

4. *Indestructibility.*—A piece of gold may be cut into parts so small as to be almost invisible. It may be dissolved by acids and made to disappear, or by intense heat it may be changed into thin vapor, and hid in the air. After all these changes have been wrought upon the gold, its particles may be again collected to form a mass like the original one without the slightest diminution in weight. Amid all the changes which we witness in the forms and qualities of bodies, not a single atom is destroyed. This property of matter, by virtue of which no particle can be destroyed, is called *indestructibility.*

5. *Elasticity.*—When an india rubber ball is pressed in the hand it is made smaller, but the moment the pressure is removed the ball springs back to its original size. The same quality is possessed in various degrees by all bodies. In such as lead and clay it is very slight, yet a ball made of either of these substances will spring back after having been for a moment compressed. On the other hand, an ivory ball, when let fall upon a marble slab, rebounds nearly to the height from which it fell, showing that the power of restitu-

tion is, in this case, almost equal to the force of compression. This property of matter, by virtue of which it restores itself to its former condition after having yielded to some force, is called *elasticity*.

This property of matter is more universal than is commonly supposed. Glass, although very brittle, is highly elastic. A glass ball will rebound from a marble slab almost as well as one of ivory. Steel is likewise hard and brittle, yet the Damascus sword could be bent double without breaking.

But should we attempt to describe all the properties of matter in detail, the time given to the study of our science would be filled with little else. The success of a student of nature depends largely upon his power to classify phenomena, and to study them in groups.

(2.) All the properties of matter may be grouped in two divisions, viz.: physical properties, of which malleability and ductility are examples; and chemical properties, such as combustibility and explosibility.

1. *Physical Properties.*—Many of the metals may be reduced to thin plates, or leaves, by hammering them. Zinc is a familiar illustration, sheets of this metal being often placed under stoves, to protect the floor from heat. This property is called *malleability*. Gold is eminently malleable: it may be beaten into leaves so thin, that a pile of eighteen hundred of them would be no thicker than a sheet of common paper.

Many substances may be also drawn into wire. Iron, copper, and brass wires are sufficiently familiar. The peculiar property by virtue of which they may be drawn into wire is called *ductility*. Glass, when heated to a bright red heat, is remarkably ductile. If a point, pulled out from the mass, be fastened to the circumference of a turning wheel, a uniform thread as fine as the finest silk may be wound at the rate of a thousand yards an hour.

Now fix the attention upon the fact that the wonderful malleability of gold, and the surprising ductility of glass, are shown *without any change in the nature of these substances.* The gold is the same material in the form of leaf as it was before it manifested its malleability. The glass in the form of thread is the identical substance which, by being drawn, manifested its ductility. All properties which, like these, a body may manifest without undergoing any change in its nature, are called *physical properties.* If now we examine those properties described in the early part of this section, we will find them all to belong to this group. Extension, impenetrability, and the rest, are properties which a body may show without any change in its nature.

2. *Chemical Properties.*—Wood, by burning, shows that it is combustible. No substance can manifest the property of combustibility except by actually taking fire, and when it burns it changes to something else.

Who, not already familiar with gunpowder, would suspect it to be so violently explosive? It can show that it is explosive, only by ceasing to be gunpowder, and becoming a mass of vapor. Properties like these, which a body can not manifest without changing its nature, are called *chemical properties.*

This classification of properties helps us to define accurately the science whose elements we are beginning to study.

(3.) Natural Philosophy is the science which treats of the physical properties of matter, and of those phenomena in which there is no change in the nature of bodies.

If now we look out upon the phenomena which nature presents, and will apply the test furnished by this definition, we may select, from among the multitude, those which it is the province of this science to explain. Thus, for example, we see the vapors rise; we see the rain drops fall. We listen with delight to the harmonies of music, and derive exquisite pleasure from the colors of the rainbow. In these phenomena, and in numerous others easily recognized by an attentive mind, we can detect changes in the form and place of bodies, but none whatever in their nature. But if we regard the more quiet, yet not less imposing phenomena of the seasons, we may discover a multitude whose discussion is, by the definition, excluded from this science. The young verdure of the spring-time changes at length to the matured foliage and ripening grains of summer. The fruits and hues of autumn, more somber, except where enlivened by the richly colored ripening leaves of the maple or the oak, soon afterward appear, only to be in turn displaced by the crisp and crackling snows of winter. These events are brought about by changes gradually taking place in the nature of substances, and the explanation of all such phenomena must be reserved for the science of chemistry.

CHAPTER I.

§ 1. THE FUNDAMENTAL IDEAS.

(4.) The fundamental ideas in natural philosophy are expressed by the words molecule, inertia, attraction and repulsion. These four ideas, when fully understood, will furnish the explanations of nearly all the phenomena of which the science treats.

1. *The Molecule.*—A molecule is a particle of matter which can not be divided *without changing its nature.* All bodies are made up of such particles. A piece of marble may be crushed and powdered until its particles are like the finest dust, yet, when seen through a microscope, they appear like angular blocks of stone, and may be still further divided. The same is true of a piece of ice. If its temperature be kept low enough while it is being crushed, every particle of the ice-powder will still be a block of ice. By applying heat, the little block is first melted, and then changed to steam, which shows that it was composed of innumerable smaller pieces. How minute must be the particles thus made absolutely invisible! Yet each one is a fragment of the original block of ice. The heat has not changed their nature. The identical particles which make up the steam, composed the drop of water

and the little piece of ice. But it is thought that these particles can not be divided *without changing their nature*, and they are called *molecules*.

All bodies are made up of molecules. The size of a body depends upon their number; its shape, upon their arrangement.

Whenever the term molecule is used, it should convey this idea, that every body of matter is made up of a multitude of little particles, which do not touch each other, and which can not be divided without changing their nature.

2. *Inertia.*—A heavy wheel requires force to put it in motion, or when in motion it requires force to stop it. It has no power to change its own condition. At rest, it would rest forever if left to itself; or once in motion it would forever move, unless acted upon by some force beyond itself. This idea, that no material body has power to change its own condition of rest or motion, is expressed by the term *inertia.*

3. *Attraction.*—When a body is not supported it falls to the ground. This familiar event illustrates the tendency of bodies to approach each other. Moreover, we have seen that bodies are composed of molecules, so small that the most powerful microscope can not reveal them, yet we must think of each as a separate body as truly as though the eye could measure its diameter. Now, by what influence are they held together? It is doubtless the same invisible force by which a body is drawn to the earth when not supported. It is a fact that all bodies, however large or small, have a tendency to approach each other. The *force* which causes this tendency is called *attraction.*

4. *Repulsion.*—If a ball of india rubber be pressed in the hand it is made smaller—its molecules are brought nearer together. When the pressure is removed they instantly spring to their former position. While springing back the molecules are evidently being thrust away from each other.

Or try the following experiment. Suspend a pith ball, or a little ball of cotton, by a fine silk thread: briskly rub a warm dry lamp chimney with a woolen cloth: bring the ball and glass together for a moment, after which it will be found that the ball will fly away from the glass, and show so strong an aversion to it that they can not be brought together. The force under whose influence bodies tend to separate is called *repulsion.*

The action of repulsion among molecules is more universal than among masses. It is illustrated by many familiar facts. If, for example, a bladder be filled with cold air, and then heated, it will burst. Repulsion drives the molecules of air apart, and pushes them through the bladder. When a drop of water is heated it becomes steam, and fills a space about 1700 times larger than before.

5. *These Four Ideas.*—Out of these four ideas may be drawn the explanation of almost all the phenomena which take place in nature. A great city, with all its various forms of architecture and machinery, is built of a few familiar substances, such as wood, and iron, and stone. This fact may excite our admiration of the intelligence and skill of man. What, then, must be our feelings when we discover that these four simple ideas are the elements out of which the sublime fabric of the universe has arisen! The whole system of material

things is simple and orderly, displaying the infinite knowledge, power, and skill of a divine Architect.

§ 2. VARIETIES OF ATTRACTION.

(5.) Attraction receives different names according to the circumstances under which it acts. Gravitation, Cohesion, Adhesion, and Capillary Force, are its most common forms.

I.—GRAVITATION.

A.—Gravitation is that form of attraction which is exerted upon all bodies, and throughout all distances. It is governed by two laws:—

1st. Its force is in proportion to the quantity of matter in the body exerting it.

2d. Its force is inversely proportional to the square of the distance through which it acts.

1. *Gravitation is universal.*—All bodies are under the influence of gravitation. The leaf, the fruit, the snow-flake, fall to the ground because they are attracted thither by gravitation. They press upon its surface because the same force continues to act after they reach the earth. No distance can outreach it, for it is the bond which holds the heavenly bodies in their orbits. Nor can any substance cut it off, or even diminish its action; for if the earth should come between the sun and moon, these two bodies would attract each other with the same degree of force.

We come now to the interesting thought that this force acts with infinite regularity and precision.

2. *The first law of gravitation.*—To illustrate this law, let us suppose two bodies, one containing twice as

much matter as the other, to attract a third. The force exerted by the first, will be twice as great as that by the other. If one body weigh nine tons and another three tons, then a third body equally distant from them will, according to the same law, receive three times as much attraction from the first as from the second.

3. *The second law of gravitation.*—To illustrate this law, suppose a body to be *twice as far* from the center, or source of attraction, at one time as at another. In the first position, the attraction will be only *one-fourth* as strong as in the second. If the distance be three times as great, the force will be one-ninth as strong. If two distances are as 3 : 4, the attractions will be to each other as 16 : 9.

Now, the weight of a body is due to the attraction of gravitation. Weight must, therefore, increase or diminish in exact accordance with the laws of gravitation. The greater the distance from the earth, the less will a body weigh. Now, distance from the earth is measured *from its center.* When, on the surface of the earth, a body is 4,000 miles from the center; suppose it were possible to carry the body to a height of 4,000 miles above the surface, its distance from the center would be *doubled*, and its weight would be reduced to *one-fourth.*

II.—COHESION.

B.—Cohesion is that form of attraction which acts between the molecules of the same body. Its power is very great, but only through insensibly small distances.

1. *Cohesion.*—Cohesive attraction holds the molecules of a body together, and enables it to keep its form

and size. A cubical block of wood remains a cube only because its molecules are held together by this force. Were it not for its action, all bodies would at once dissolve into their ultimate molecules, and vanish.

2. *Its power.*—The strength of cohesion is often very great. The molecules of a piece of iron are so strongly bound by it, that a weight of 500 lbs. may be lifted by means of a wire one-tenth of an inch in diameter. Even a strip of paper is not easily broken by a force acting exactly in the direction of its length.

3. *It acts through insensible distances.*—The distance through which cohesion can act is quite too small to be measured. Let the parts of a body be separated, and the strength of the giant is gone.

When a body is broken its parts can be made to cohere again only with great difficulty. In a few soft bodies, like wax, a slight pressure will force the molecules near enough together for cohesion to take hold of them; in others the pressure required is much greater, while in the majority of substances it is so great as to be practically impossible.

The smith unites two pieces of iron by *welding.* He softens the iron by heat, then puts the two pieces together and unites them by the heavy blows of his sledge. Now, what he does is simply to push the yielding molecules of the two pieces of iron into very close contact; this done, cohesion grasps them, and the two pieces become one.

III.—ADHESION.

C.—Adhesion is that form of attraction which acts between molecules of different bodies without changing their nature.

If, for example, the hand be plunged into water it comes out covered with a thin film of the fluid; it may be immersed in alcohol with the same effect. In these cases the fluids are held to the hand by *adhesion*. The hand may be withdrawn from a bath of mercury without retaining a particle of that substance, because the adhesion is too feeble to lift the fluid.

This force, like cohesion, acts only through distances too small to be measured: unlike cohesion it acts between molecules of different kinds of matter. The value of glue and cement is due to the powerful adhesion which acts between them and the surfaces of solid bodies which they bind together.

IV.—CAPILLARY FORCE.

D.—Capillary Force is the adhesion of a liquid to a solid which is partly immersed in it. It generally causes an elevation or a depression of the liquid along the sides of the solid. It also causes a liquid to penetrate a porous solid. It is governed by two laws:—

1st. The heights to which a liquid rises in different tubes of the same material are inversely proportional to the diameters of the tubes.

2d. The height to which a liquid rises between parallel plates is one-half the distance it will rise in a tube whose diameter is equal to the distance between the plates.

1. *Capillary Force.*—If small glass tubes be inserted in a vessel of water, it will be seen that the fluid instantly springs upward and remains at rest in the tubes considerably above its general level. (See Fig. 1.) Along

the outside surface of the tubes the water also climbs to a little height.

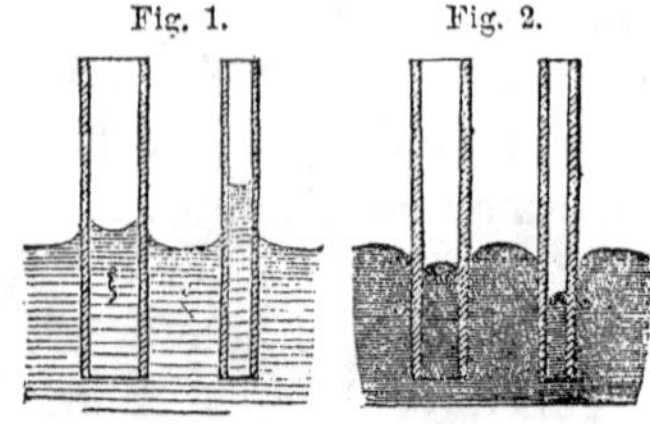
Fig. 1. Fig. 2.

If tubes be inserted in a vessel of mercury, this fluid will be pushed down. (Fig. 2.) The mercury inside the tubes will be considerably below the general level, while the fluid against the outside is also depressed. Here are two well marked cases of capillary action.

Now, when a piece of glass is plunged into water it comes out wet, but when plunged into mercury it comes out as free from the liquid as when it entered, and by repeated experiments it is shown that all liquids which will wet the sides of the tube will be *lifted*, while those which will not, will be *pushed down*.

2. *It causes liquids to penetrate porous solids.*—An easy experiment strikingly illustrates this action. Take a common bottle, eight or ten inches high, and wrap it in a sheet of white blotting-paper, whose edges must be secured by a bit of wax. Place the bottle, now prepared, upon a dinner-plate. Pour water upon the plate to cover the lower edge of the paper, and immediately the fluid will be seen rapidly climbing the sides of the bottle, which it will not cease to do until it has reached the top. The beauty of the experiment is enhanced by filling the bottle with some highly colored liquid.

The rise of the water is due to the attraction between its particles and those of the paper and glass. This force, acting downward from each particle of the paper through the definite but imperceptible distance to the one below it, lifts a particle of water. The next particle of paper above, then lifts it higher. Indeed, the successive particles of paper upward, are the successive steps of a ladder, up which the water is impelled by capillary force.

Numerous familiar facts are explained by this experiment. Oil is carried up the lamp wick to supply the flame with fuel. By a similar action, water is distributed through loose soils to keep them moist and fertile. So, too, in a great degree, the sap of plants and trees is carried to their summits, and even in the animal system the circulation of blood through the minute blood-vessels is materially aided by capillary action.

3. *The first law.*—In figure 1, the water is represented as being lifted to different heights in the different tubes. The height to which any fluid rises depends upon the size of the tube. If the diameter of one tube be just one-half that of another, water will invariably rise in it twice as far. If the diameters of two tubes have the ratio of 4 : 3, then the water will rise in them to heights whose ratio is 3 : 4. Or, in other words, the heights are inversely as the diameters of the tubes.

4. *The second law.*—If a plate of glass be inserted in water the liquid will rise a little distance against its sides. If two parallel plates be inserted near together, the water will rise between them, and by varying their distance from each other it may be shown that *the height to which the liquid rises is inversely proportional to the distance between the plates.*

But if we compare the elevations which take place between the plates with the height to which the same liquid rises in tubes whose diameters are equal to the distance between the plates, we discover that in all cases it is just *one-half.*

(6.) We need not suppose that gravitation, cohesion, adhesion, and capillary force are so many different kinds of force. They should be regarded as but different manifestations of a single influence.

One can not, it would seem, study the phenomena thus far briefly sketched without being impressed with the variety of the phases and effects of attraction, yet we do well to regard all these varied effects as but the different ways in which a single influence manifests itself. The differences between gravitation, cohesion, and the other forms of attraction that have been named, are apparent, not real. When bodies are separated by sensible distances, the attraction between them is called gravitation, without regard to their size; but when the bodies become very small, and the distances very minute, the force is called cohesion. Does *distance* alone, then, change the *nature* of attraction? Where shall the line be drawn at which the change occurs? Equally unreal are the differences between the other members of this group of forces, so that, looking behind the veil of appearances, we are, upon the very threshold of science, permitted to catch a glimpse of the sublime simplicity which everywhere reigns in the works of nature, and which it is the glory of scientific study to reveal.

But shall we ask what is this single power whose effects are so varied and imposing? We name it attraction. Who need inquire further? The little bird that vainly beats his head against the cage bars, affords a warning to the man of science who would attempt to search for this, which God has hidden.

Problems Illustrating the Laws of Attraction.

1. With how many times greater force will a body be attracted by a mass of iron weighing 9 tons, than by a block of stone weighing 3 tons, when both are at the same distance from it? *Ans.* 3.

2. Two lead balls, one weighing 5 ozs. and the other 12 ozs., are hanging at a distance of 10 ft. from a third; what relative degrees of force do they exert upon it?

3. One ball of lead attracts another through a distance of 10 ft., with a force of 8 lbs.; what force would it exert if placed at a distance of 20 ft? *Ans.* 2 lbs.

4. A body is at one time 50 ft., and at another 75 ft., from a mass of rock; what are the relative forces exerted upon it in the two positions? *Ans.* 9 : 4.

5. Two bodies, one weighing 6 lbs. and the other 9 lbs., are attracting a third. The first is at a distance of 25 ft., the second of 50 ft.; what relative attractions do they exert? *Ans.* 24 : 9.

6. At the surface of the earth a body weighs 10 lbs.; what would it weigh if carried to a height of 5 miles above the surface? *Ans.* 9.97 lbs.

7. A glass tube $\frac{1}{100}$ inch in diameter raises water by capillary force about 4 inches; how high will water rise in a tube $\frac{1}{10}$ inch diameter? *Ans.* $\frac{4}{10}$ in.

8. How high will water rise between two parallel plates $\frac{1}{10}$ inch apart? *Ans.* $\frac{2}{10}$ in.

9. If between parallel plates $\frac{1}{100}$ inch apart, water rises two inches; how high will it rise when the plates are $\frac{1}{20}$ inch apart? *Ans.* $\frac{2}{5}$ in.

CHAPTER II.

OF THE THREE PHYSICAL FORMS OF MATTER. INTRODUCTION.

APPLICATION OF THE FUNDAMENTAL IDEAS.

(7.) Read (4). Attraction and repulsion acting upon the molecules of bodies, produce the three physical forms of matter: solid, liquid, and gaseous.

Between the molecules of every body, two sets of forces, attraction and repulsion, are continually struggling. Just in proportion as one or the other prevails, the body will be a solid, a liquid, or a gas. In a solid body attraction prevails, and its molecules are firmly bound together. In a liquid body the attraction is almost equaled by the repulsion, and the molecules are left free to move easily among themselves. In a gaseous body the repulsion exceeds the attraction, and the molecules are driven away from each other to the greatest possible distance. The solid rock, the mobile water, and the rushing air, are types of these three grand divisions to which all bodies belong.

The attraction and repulsion among the molecules of bodies are called *molecular forces*. Cohesion, adhesion, and capillary force are molecular attractions; the force of heat is a molecular repulsion.

Numerous and familiar changes of form are due to the action of heat. Ice, for example, when heated, becomes water, and water, when heated still more, rises in vapor to form the floating clouds. Or suppose the action to be reversed. Imparting their heat to other bodies, the clouds are changed to water, and water, again to solid ice and feathery snow.

Imitating nature, we may to a limited extent, by the use of heat, change the form of various bodies, and numerous arts of life spring from the application of this power. By the repulsive force of heat the metallic ores are melted, and the useful metals obtained. By the same force iron is liquefied, that it may be molded into requisite forms of strength, of beauty, or of use, demanded in the arts. The expansive force of steam is but the repulsive force of heat.

§ 1. OF THE CHARACTERISTIC PROPERTIES OF SOLID BODIES.

(8.) The characteristic properties of solid bodies are hardness, tenacity, malleability, ductility, and crystalline form.

1. *Hardness.*—The particles of solid bodies are held together by cohesion, much more firmly in some than in others. Those in which they are held with the greatest force, will most successfully resist the pressure of others. By the term *hardness*, we refer to that property of solids which enables them to resist any action which tends to wear or scratch their particles away.

Hardness does not imply strength. A piece of glass will scratch an iron hammer, which proves it to be harder than iron, yet glass is very fragile, easily broken by the stroke of soft wood; indeed, by almost any thing that can inflict a blow.

Neither does hardness imply density. The diamond is the hardest of substances, while gold is so soft as to be easily cut with a knife; yet gold is four times as dense as the diamond. Mercury is a fluid, and, of course, has no hardness, yet it is nearly twice as dense as the hardest steel.

The process called tempering or annealing, consists in regulating the hardness of a body by the action of heat. Steel, when in its hardest condition, is too brittle to be used in the arts; but by heating it to a temperature determined by the use to be made of it, and then slowly cooling it, the steel may receive any degree of hardness desirable. It may be made almost as soft as soft iron, or it may become nearly as hard as the diamond.

2. *Tenacity.*—When a rod of iron is stretched in the direction of its length, it will be found that great force is required to pull it apart. The property, in virtue of which bodies resist a force acting in the direction of their length, is called *tenacity.*

The metals are more tenacious than other solids, and among metals, iron in the form of cast-steel, stands at the head of the list. A rod of cast-steel, the end of which has an area of one square inch, will support a weight of 134,256 pounds.

It has been found by experiment that the tenacity of a bar is in proportion to the area of its cross section, and entirely independent of its length.

It has also been shown that the tenacity of a metal is greatly increased by drawing it into wire. The cables of suspension bridges are, for this reason, made of fine iron wire twisted together.

3. *Malleability.*—The particles of many solid bodies may be displaced without overcoming their cohesion. By the blows of a hammer, the molecules of many metals may be shifted about, without breaking them apart, until the bodies are reduced to the form of thin plates or leaves. By passing the metal between the rollers of a rolling-mill, the great pressure exerted will produce the same effect. This property, in virtue of which a body may be hammered or rolled out into thin leaves or plates, is called *malleability.*

This property is possessed in a high degree by many of the metals. Under the hammer, lead is the most malleable of the useful metals; tin stands second, and gold third on the list. In the rolling-mill, gold is the most malleable, silver is second, copper third, while tin stands in the fourth place on the list. (Cooke's Chemical Physics, p. 207.)

4. *Ductility.*—If, instead of being reduced to thin plates, the substance may be drawn into wire, the property thus shown is called *ductility.* This property is closely allied to malleability, but metals do not possess both in an equal degree. Platinum, for example, which is seventh on the list of malleable metals, stands first on the list of those which are ductile. This metal has been drawn into wire finer than a spider's thread.

5. *Crystalline Form.*—The attraction among the molecules has not brought them together at random, nor in disorder. A flake of snow, when seen through a microscope, is found to be as symmetrically formed as a swan's feather; and water frozen on the window panes in winter shows a beautiful variety of tree-like forms. These definite and regular forms in which solid substances occur are called *crystals*, and any process by

which they may be obtained is called a process of *crystallization.*

In the formation of solid bodies their tendency to take a crystalline form is almost universal. The same substance generally takes the same form, but in different substances the shape of crystals may be wonderfully unlike. Lead, in its most common ore, called *galena*, is found crystallized in cubes. Specimens of these cubes are often found as perfect as could be chiseled by an artist.

But the larger number of solid bodies around us do not appear to have these definite crystalline forms. They have been made solid under circumstances which did not allow the molecular forces to act freely. In many cases, however, if we break open a body whose external form is not regular, we may discover that it is, after all, a crystallized body, by noticing that it is made up of multitudes of small crystals, very closely packed together. This is true of many rocks.

Even when no indication of a crystalline structure can be seen, the substance can often be made to assume it by some artificial process. The best method is to dissolve the solid in water or some other liquid, and allow the solution to stand in a quiet place where it may evaporate slowly. Common salt and alum are substances which readily and beautifully illustrate this process. The more slowly the water evaporates, the more perfect will the crystals be. (See Cooke's Chemical Physics, pp. 119 to 185.)

§ 2. OF THE CHARACTERISTIC PROPERTIES OF LIQUID BODIES.

(9.) Liquids have elasticity and some other properties in common with solid bodies. But since the attraction and repulsion among their molecules are very nearly equal, we find that mobility is their characteristic property.

1. *Elasticity.*—When submitted to pressure liquids are compressed, and when the pressure is removed they instantly spring back to their original volume. It requires a very great force, however, to compress a liquid in the least degree; so great, that until improved means of experiment were contrived, liquids were thought to be incompressible. Water, at a freezing temperature, when pressed by a force of 15 lbs. to the square inch, is condensed only .0000503 of its volume. (See Cooke's Chemical Physics, pp. 215 to 218.)

The force with which a liquid springs back to its former size after being compressed, is exactly equal to the force which compressed it; it is, for this reason, said to be *perfectly elastic.*

2. *Attraction and repulsion nearly equal.*—That the attractive and repulsive forces among the molecules of a liquid are not exactly equal may be shown by a pretty experiment.

To one end of a scale-beam (Fig. 3) a disk of brass is suspended, and accurately balanced by weights in the opposite scale pan. Now let the disk be brought to rest upon the surface of water in a vessel, and it will be held there with considerable force. If the disk be two inches in diameter, weights equal to 200 grs. may be piled upon the opposite pan

Fig. 3.

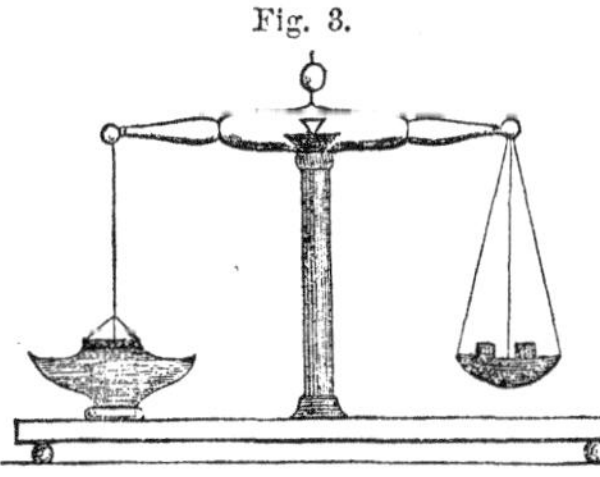

before it will be torn from the water. Now, notice that a film of water still adheres to the disk, having been torn away from the water beneath it. The 200 grains weight have simply overcome the *cohesion* of the water. We thus learn that the attraction is a trifle stronger than the repulsion.

But the attractive and repulsive forces are *nearly* balanced, and if we now remember that water consists of molecules, it is not more difficult to see that there must be freedom of motion among them, than it is to see that a number of smooth balls will roll easily upon each other.

3. *Mobility.*—To illustrate the mobility of water, and its cause, let the following simple experiment be tried.

Take three glass goblets: fill one with small marbles, one with fine shot, and the third with water. After putting a dinner-plate over each goblet, they may be inverted without spilling their contents. Now, lift the first, and the marbles will roll out upon the plate. Lift the second, and the shot roll out in the same way. A person at a distance will not be able to see the separate shot, but will see their motion, and know it to be caused in exactly the same way as the motion of the marbles, which could be seen distinctly. Now lift the third goblet, and the water spreads out upon the plate exactly as did the marbles and the shot. The molecules of water are balls infinitely smaller than shot; but, while the most powerful microscope fails to reveal them, the mind can see them, so small, so round and smooth, that they roll and glide among themselves with the greatest freedom.

The phenomena peculiar to liquid bodies depend

chiefly upon the mobility of their particles. The phenomena of liquids *at rest* must be now considered: those of liquids *in motion* must be reserved for a future chapter.

§ 3. OF LIQUIDS AT REST.

(10.) At any point inside of a body of liquid there is equal pressure from all directions.

Hence a fluid will rest only when its upper surface is level. And the level surface of a large body of water is convex.

1. *Liquids press in all directions.*—In order to see that, because the particles of a liquid are free to move, they must be exerting pressure in all directions, we will suppose a number of very smooth balls to be arranged as in Fig. 4. The weight of the ball A will be a downward pressure upon the balls B and C. These, being free to move, will be pushed aside. The ball B, moving toward the left, will push between the balls D and E, while the ball E, moving upward, will exert an upward pressure.

Fig. 4.

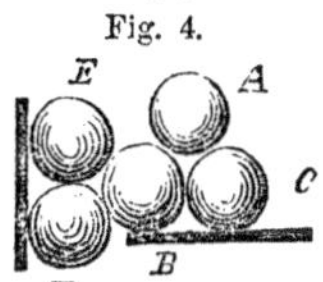

Just so the small molecules of a liquid are exerting pressure downward, upward, and laterally; and, moreover, if the liquid be at rest, every point in it must be pressed *equally* in all these directions.

An experiment may help to illustrate this principle. If a disk of metal be held in the middle of a jar of water, it is easy to see that it must be pressed downward by the weight of the water just above it; but it may not be so clear that it is pushed up by an equal force. Taking a lamp chimney, and putting the string handle of the disk C, Fig. 5, through it, hold the disk tightly against the lower end of the tube until it is pushed down to the middle of the water. Now loosen the string: the heavy disk does not sink, but remains tightly pressed *upward* against the tube by the water. If water be allowed to enter the tube it will press down upon the disk, and when it has filled the tube almost to a level with the water outside, then the disk falls, suggesting that the upward and downward pressures are equal.

Fig. 5.

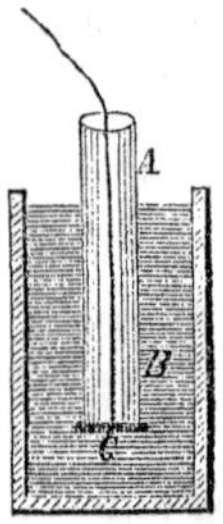

Numerous simple experiments might be given to illustrate this important principle: one other must suffice. Glass is eminently brittle. It may be blown into sheets as thin as the finest paper cambric. In this condition, the weight of a few grains resting upon it in the air would crush it. Yet, placed near the bottom of the deepest cistern, it will support the weight of all the water above it, and remain unbroken. This could not happen, if the pressure of the water upon it was not equal from all directions.

2. *The surface of water at rest is level.*—The truth of this principle may be seen by attentively examining Fig. 6, which represents a section of a vessel containing water, the surface of which has for the moment been thrown into the position indicated by the line A E. Refer to any two points in the water, as *h* and K. We see that the downward pressure at *h*, would be the weight of the water above that point—a column *m h*. But the pressure at that point is equal in *all* directions,

Fig. 6.

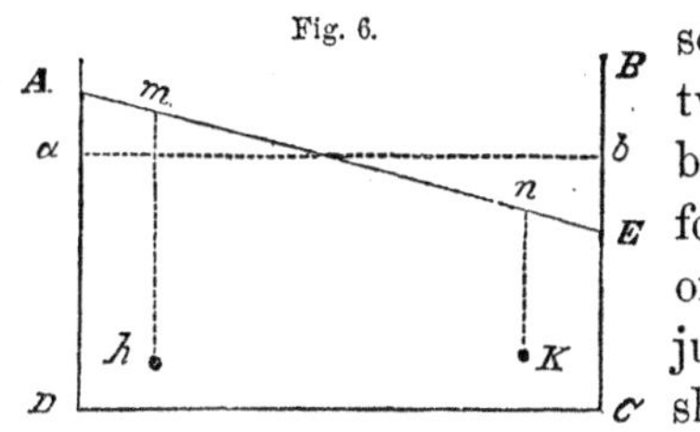

so that the water between h and K would be pressed toward K by a force equal to the weight of the column $m\ h$. In just the same way we may show that at K, the water is being pressed toward h by a force equal to the weight of the column n K. The column $m\ h$ is greater than n K, and since the water is *free to move* it will yield to the greater pressure, and go toward K until the two forces are equal. The two forces will be equal only when m and n are in the same level surface, $a\ b$.

3. *But a level surface is convex.*—The surface of water will be at rest when the force of gravitation acts upon all points of it alike. That the attraction of the earth may be equal on all points, they must be equally distant from the center of the earth. To be at the same distance from the center of the earth they must form a curved surface. In case of large bodies of water, of the oceans for example, the convexity can be seen. It is shown by the ancient observation that the topmast of an approaching ship is the part first seen from port.

(11.) Since the surface of water at rest must be level, we infer that water confined in pipes or close channels will always rise as high as the source from which it comes.

Upon this principle cities are often supplied with water.

The same principle explains the phenomena of springs and artesian wells.

1. *Water in pipes will rise as high as its source.*—

If into one arm of a bent tube we pour water, it will flow around into the other, until it stands at the same height in both. No matter what may be the shape of the vessel, the surface of the liquid it holds must be just as high in one part of it as in another, and a pipe leading from a vessel is a part of the vessel which holds the water.

2. *The supply of water to cities.*—A pipe leading from a reservoir of water on a hill outside a city, may be buried in the ground, passed down the hill-side, and through the streets, and be provided with branches leading into cisterns in every dwelling. Unless these cisterns are higher than the water in the distant reservoir, the water will flow down the hill-side, through the streets and up the branches into the dwellings, and supply them all with water. Many cities are, in this way, conveniently supplied with abundance of water, for private dwellings not only but for public fountains and manufacturing purposes.

3. *Springs.*—The rocks which compose the earth are arranged in layers, called strata, which are generally more or less oblique as represented in Fig. 7. Some of these strata will allow water to soak through them; others will not. In the figure the dotted portions *a a a* indicate the porous strata.

Now, water falling on the surface of the earth at *c*, will settle through the loose or porous material until it reaches the rock, which it can not penetrate. Flowing along the surface of this rock, it will issue from the hill-side at S, and thus form a *spring.*

4. *Artesian wells.*—Again, the water, falling upon the surface and passing through other porous layers, comes in contact with a rock which it can not penetrate,

and flows along its surface. The basin-shaped part, *a a*, of the porous layer, would thus in time become filled

Fig. 7.

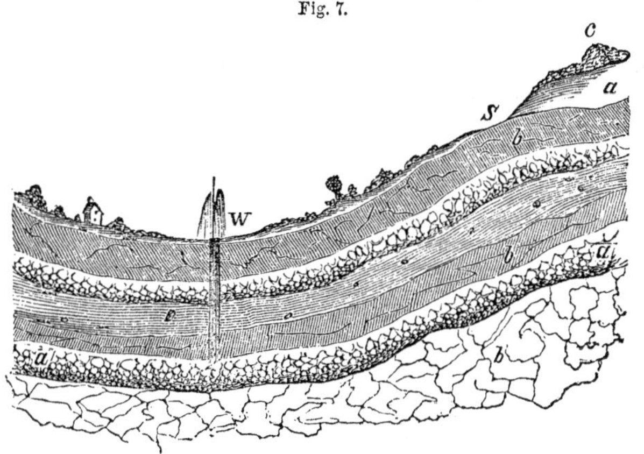

with water; indeed the entire layer reaching to the surface of the earth in both directions might thus be filled. If, then, a well at W be sunk through the mass down to this saturated layer, the water will rise in the well, sometimes to the surface of the ground above, and often spout in jets many feet above it.

Such wells are often bored to very great depths and are called artesian wells. One of these wells was bored in Louisville, Kentucky, to the depth of 2,086 feet. Another in St. Louis has a depth of 2,199 feet. The supply of water furnished is often very abundant. The famous Grenelle well, in Paris, yields daily 600,000 gallons.

(12.) The pressure of a liquid on the bottom of the vessel which holds it is independent of the shape of the vessel. It depends on the depth of the liquid, and equals the weight of a column whose base is the base of the vessel, and whose height is the depth of the liquid in it.

1. *The pressure is independent of the shape of the vessel.*—This may be proved by experiment. The essential parts of an apparatus for this purpose are represented in Fig. 8. A glass tube, A B, bent twice at right angles, contains mercury. The height of the mercury in one arm is shown by a graduated scale, and to the other arm vessels of various forms and heights may be attached. When a vessel, G, is filled with water, the fluid presses upon the mercury at A, and pushes it up in the arm C D; the height to which it rises being shown by the graduated scale. Now let the vessel be removed, and another, in the form shown at E, be put in its place. If water be poured into this vessel until it stands as high as it did in the other, the mercury will be seen to rise in C D to the same point as before. Vessels of various other forms may be used, but if all are of the same height the water which fills them will push the mercury to the same point on the scale. We infer that the pressure of a fluid downward is quite independent of the shape of the vessel and the quantity of fluid.

Fig. 8.

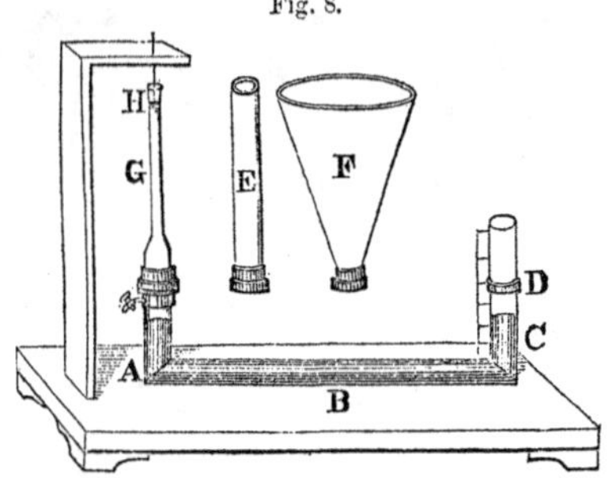

2. *The pressure depends on the depth of the liquid.*—If a tube twice as high as the vessel E, in Fig. 8, be

used and filled with water, the mercury will be seen to rise just twice as far as when the other vessels were employed, and by repeated experiment it is seen that the pressure is in proportion to the height of the column of water which exerts it. (See Cooke's Chemical Physics, p. 223).

3. *To calculate the pressure.*—If the pressure depends only on the size of the base and the height of the column, then it must equal the *weight* of a column whose base is the base of the vessel, and whose height is the depth of the liquid. Now, *one cubic foot of water weighs* 62½ lbs., and if the number of cubic feet of water which exerts the pressure be multiplied by 62½, the amount of pressure in pounds will be obtained. Thus, suppose a vessel, represented by E F C D, in Fig. 9, to be full of water: the pressure on its bottom is the weight of the column A B C D. Let the area of the bottom be 3 square feet, and the depth of the water be 8 feet; then 24 cubic feet of water exerts the pressure, and 24 × 62½ lbs., or 1,500 lbs. is the pressure exerted.

Fig. 9.

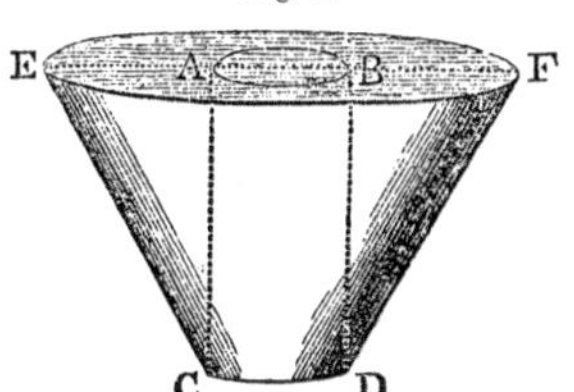

Now, since the pressure is equal in all directions, we may obtain the amount of pressure against *any portion of surface* either in the bottom or sides of the vessel, by getting the weight of a column of water whose base is the surface pressed upon, and whose height is the depth of the water *to the middle point* of that surface. For example, suppose we would know how much pressure is borne by one square foot of the *side* of a vessel at a depth of ten feet below the surface of the water.

We must understand that ten feet is the distance from the top of the water to the *middle point* of the square foot; then the pressure will be the weight of a column of water whose base is one square foot and whose height is ten feet. Such a column will contain 10 cub. ft. of water, and its weight will be $10 \times 62\frac{1}{2}$ lbs.

(13.) A solid body when immersed in a fluid, is pushed upward by it with a force equal to the weight of the fluid it displaces.

It follows from this principle, 1st. That a solid body, lighter than water, will sink far enough to displace water whose weight is equal to its own. 2d. That a solid body, heavier than water, will weigh less in water than in air, the difference being the weight of the water displaced by it.

1. *Solid bodies in water are pressed upward.*—If, for example, a piece of wood be pushed down into a vessel of water we find it struggling to rise to the surface. It is pressed upward by the water under it, and considerable force of the hand is required to keep it down. Or if a stone be suspended in water it feels lighter than when in air; the water under it pushes upward against it, and thus supports a part of its weight.

Fig. 10.

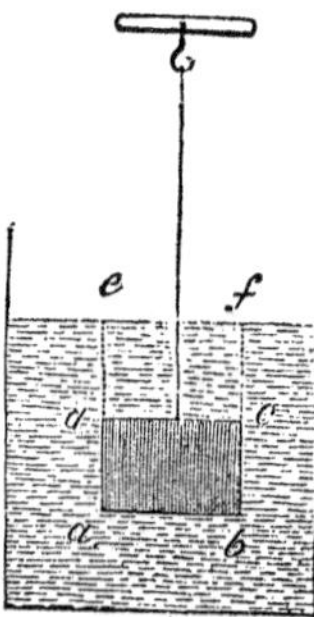

2. *With force equal to the weight of water displaced.*—Now, suppose a block of marble suspended in a vessel of water (Fig 10). The upward pressure against its lower surface, *a b*, is equal to the downward pressure of the water at that depth, and this downward pressure is equal to the

weight of the column of water, *e f b a*. Now the column of water, *e f c d*, is sustained by a part of this upward pressure, and the rest of it is exerted upon the marble. To sustain the column, *e f c d*, requires an upward pressure equal to its weight, and hence there is left a pressure against the surface, *a b*, equal to the weight of a column of water, *a b c d*, but this water is displaced by the marble. The upward pressure against the block is, therefore, equal to the weight of the fluid displaced.

We owe the discovery of this important principle to Archimedes, one of the most eminent philosophers of antiquity, and to this day it is called the *principle of Archimedes*. Its applications are numerous. It helps the chemist to distinguish one substance from another, and the merchant, often, to judge of the purity and value of his merchandise. In any case it enables the inquirer to determine the size or volume of a solid body, however irregular, and it has, moreover, led to valuable improvements in marine architecture and in other arts.

3. *If the solid is lighter than water.*—The weight of the water displaced by a block of wood will just equal the weight of the wood itself. A pound of wood will displace a pound of water, but a pound of wood is larger than a pound of water, so that only part of the wood will be immersed. A tin basin and a wooden bowl of the same size, will displace an equal volume of water, if the walls of the basin are thin enough, so that the two bodies have the *same weight*. Upon this principle iron ships are built. An iron ship will sink no farther than one of wood of the same size, provided the walls of iron are so thin that the two ships shall be of the same weight.

4. *If the solid is heavier than water.*—If a solid be heavier than water, the upward pressure of the fluid can support only a part of its weight. The weight supported will be the weight of the water which the solid displaces. Thus, for example, a piece of marble which weighs 10 ozs. in air, will be found to weigh only 6.3 ozs. in water. The upward pressure of the water is equal to 3.7 ozs., and this is the weight of the water which the marble displaces, and whose bulk is, of course, just equal to the bulk of the marble.

(14.) The specific gravity of a substance is its weight compared to the weight of an equal bulk of some other body taken as a standard.

To obtain it, different methods must be taken, according as the body is a *gas*, a *liquid* or a *solid.*

1. *Specific Gravity.*—The specific gravity of a substance shows how many times heavier it is than an equal bulk of some other body. The standards used are *water* and *air;* water for all solid and liquid bodies, and air for all gases. Then, when we say, for instance, that the specific gravity of gold is 19, we only mean that a cubic inch of gold will weigh 19 times as much as a cubic inch of water. The specific gravity of oxygen gas is 1.106: that is to say, a cubic inch of oxygen gas will weigh 1.106 as much as a cubic inch of air. The following simple rule must evidently cover all cases of getting specific gravity:—*Divide the weight of the body by the weight of an equal bulk of the standard.*

I.—OF GASES.

A.—To obtain the specific gravity of a gas, divide

the weight of a convenient portion of it by the weight of an equal portion of air.

To get the weight of equal portions of gases is, however, a difficult process, requiring many precautions. Without trying to give the details of the operation (see Cooke's Chem. Phys., pp. 93 and 667), we may describe it in general terms. A glass globe is first weighed when full of air. The air is then taken out of it by means of an air-pump, and the globe is again weighed: the difference in these weights is the weight of the globe full of air. The globe is then filled with the gas whose specific gravity is desired, and again weighed: the difference between this weight and that of the empty globe, is the weight of the globe full of gas. The specific gravity is obtained from these weights of equal volumes of gas and air.

II.—OF LIQUIDS.

B.—The specific gravity of a liquid may be obtained in various ways. We may notice—

1. By direct weighing.
2. By an instrument called the hydrometer.
3. By the use of a solid bulb.

1. *By direct weighing.*—The most direct method of getting the specific gravity of a liquid, is to weigh equal quantities of it and water, and then divide the weight of the liquid by that of the water. To facilitate the operation, "specific gravity bottles" are made, which hold just 1,000 grains of pure water. The weight of the bottle being known, a single operation with the

balance will give the weight of the liquid, and then its specific gravity may be speedily calculated.

2. *By the hydrometer.*—A common form of this instrument is represented in Fig. 11. It consists of a glass tube, with two bulbs near its lower end. The tube and upper bulb are full of air, which renders the instrument lighter than water. The lower and smaller bulb contains shot enough to keep the instrument in an erect position, when placed in a liquid, as shown in the figure. A graduated scale is fixed to the stem, to indicate the depth to which the instrument sinks in different liquids.

Fig. 11.

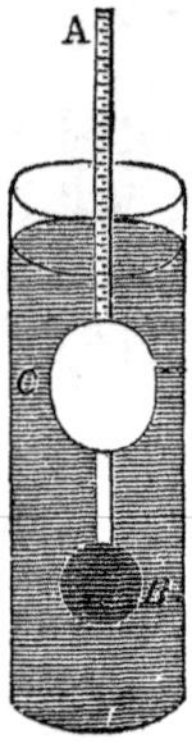

The action of this instrument can be readily explained by means of a piece of wood, several inches long and an inch square, having its lower end loaded with wire. If this be put into a vessel of water, it will sink to a certain depth, and remain upright. If it sinks 10 inches, then 10 cubic inches of water are displaced by it. If now the instrument be put into a vessel of alcohol, it will sink deeper, suppose it be 12 inches; then 12 cubic inches of alcohol are displaced. But, according to the principle of Archimedes, the fluid displaced is equal in weight to the body floating in it [see (13), 1 and 2]. Hence 10 cubic inches of water have the same weight as 12 cubic inches of alcohol, or alcohol is $\frac{10}{12}$ as heavy as water. Its specific gravity is, therefore, $\frac{10}{12}=.833+$.

Making the instrument of glass, and giving it the form seen in Fig. 11, renders it more convenient, but does not alter the principle on which it acts.

The graduation of the scale is arbitrary, and varies in different forms of the instrument. The zero usually marks the point to which the hydrometer sinks in pure water, and the degrees above and below show how far the instrument may sink in liquids respectively lighter and heavier than water.

3. *By the use of a bulb.*—According to the principle of Archimedes, a heavy bulb of glass, or other convenient substance, when weighed in any liquid, will lose a part of its weight just equal to the weight of an equal bulk of that liquid. Hence, weigh a bulb of glass in air, afterward in water, and then in the liquid whose specific gravity is desired. The *losses* of weight it sustains will be the weights of equal bulks of the two liquids, and from these weights the specific gravity may be obtained,

To illustrate this method, suppose the specific gravity of alcohol is to be found. A bulb of glass, weighed in air and then in water, is found to lose 325 grs. Its loss in alcohol is found to be 257 grs. Then $\frac{257}{325}$=.79+ is the specific gravity of the alcohol.

III.—OF SOLIDS.

C.—There are two important cases of common occurrence:—

1. The solid is heavier than water.
2. The solid is lighter than water.

1. *Of a solid heavier than water.*—Divide the weight of the body in air, by its loss of weight in water. The principle of Archimedes explains this rule.

Thus the weight of marble [see (13), 4] in air being 10 ozs., and in water being 6.3 ozs., it is clear that the difference, 3.7, is the weight of a bulk of water equal to the size of the marble. Then, $\frac{10}{3.7} = 2.7$ is the specific gravity of this solid.

The experiment is conducted in the following manner. Let the specific gravity of iron be desired. A fragment of iron of convenient size is hung from the bottom of one scale pan of a balance, and weighed. It is then immersed in a vessel of water (see Fig. 12), and its weight again determined.

Fig. 12.

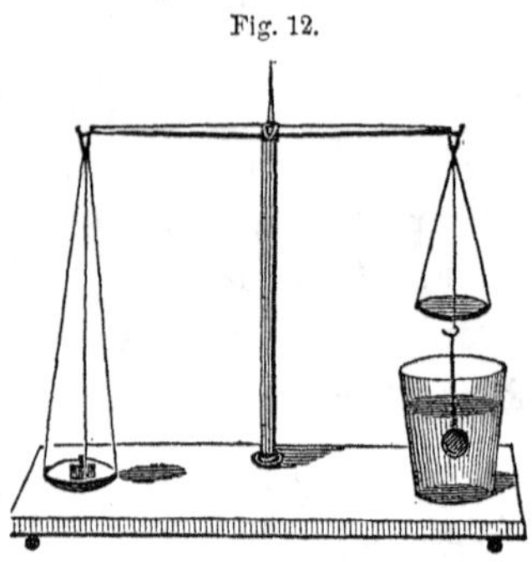

Now, suppose the iron weighs, in air, 360 grs., and in water, 313.85 grs. Then $360 - 313.85 = 46.15$ grs. is the weight of an equal bulk of water. And $\frac{360}{46.15} = 7.8$ is the specific gravity of the iron.

2. *Of a solid lighter than water.*—If the solid be lighter than water, the operation is more complex. If the light body be *compelled* to sink in water by fastening to it some heavier body, their loss of weight will represent the upward pressure of the water upon them both. If the heavy body alone be weighed in water, its loss will represent the upward pressure against it. Now, if the upward pressure against the heavy body be subtracted from the upward pressure upon both, the difference must represent the upward pressure against the light body alone, and hence, the weight of a quantity of water equal to its bulk.

To illustrate this operation, suppose a body weighed, in air, 200 grs. When attached to a piece of lead, both weighed 1,936 grs. in air, and 1,460 grs. in water, suffering a loss of 476 grs. The lead itself, when weighed in water, lost 152 grs. The upward pressure against the light body alone must then have been, $476 - 152 = 324$ grs. Then, 200 grs., the weight of the light body in air, divided by 324 grs., the weight of an equal bulk of water, is the specific gravity desired.

In the following table the specific gravity of various substances are arranged for reference.

I.—OF GASES, AT 32° F. BAROMETER, 30 INCHES.

Names.	Sp. gr.	Names.	Sp. gr.
Air	1.000	Nitrogen	0.972
Oxygen	1.106	Carbonic Acid	1.529
Hydrogen	0.0691	Olefiant Gas	0.978

II.—OF LIQUIDS, AT 39° F.

Names.	Sp. gr.	Names.	Sp. gr.
Water (distilled)	1.000	Ether	0.715
Sea Water	1.026	Naphtha	0.847
Milk	1.030	Oil Turpentine	0.869
Alcohol (absolute)	0.792	Wine of Burgundy	0.991
Olive Oil	0.915	Mercury (32°F.)	13.596

III.—OF SOLIDS, AT 39° F.

Names.	Sp. gr.	Names.	Sp. gr.
Platinum	21.5	Silver (cast)	10.47
Gold (cast)	19.26	Diamond	3.50
Iron "	7.2	Marble	2.70
Steel	7.8	Ivor	1 92
Lead (cast)	11.3	Ice	0.93
Copper "	8.8	Pine wood	0.66

(15.) If an external pressure be exerted upon any portion of the surface of a liquid, the same amount of pressure will be transmitted equally in every direction.

This is true, whatever may be the form of the vessel which contains the liquid. This is the principle applied in the hydrostatic press.

1. *The equal transmission of pressure.*—To illustrate the principle stated above, let a vessel, represented in section by Fig. 13, be quite filled with water. In the sides of the vessel are several apertures, A, B, C, D, and E, closed with movable pistons. Let the area of each piston be 1 sqr. in., and suppose a weight of two pounds be placed upon the piston A. It will be found that a force of two pounds will be exerted against each of the other pistons. Thus E will be pushed upward by a force of two pounds, while B and D will at the same time be pushed in opposite directions, each with a force of just two pounds. No matter how numerous these pistons may be, nor in what direction they may be inserted, each will be found exerting a two-pound pressure under the influence of a force of two pounds acting at A. *Every square inch in the entire surface of the vessel* will receive a pressure of two pounds.

Fig. 13.

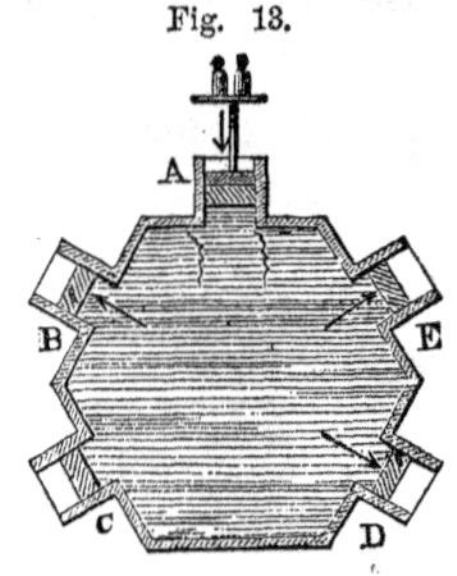

2. *The shape of the vessel makes no difference.*—The vessel may be of any shape whatever, and an equal pressure will be received on every square inch of its surface whenever an external force is applied. We will suppose, for illustration, that the vessel is a bent tube in the form of the letter U.

A pressure may be exerted upon the water in one arm, by forcing the breath into the open end of the tube. The liquid will go down in that arm and rise in the other. The pressure of the breath is *downward* in one arm; it is *lateral* through the bend, and *upward* in the other arm. Moreover, these pressures are all exactly equal.

Fig. 14.

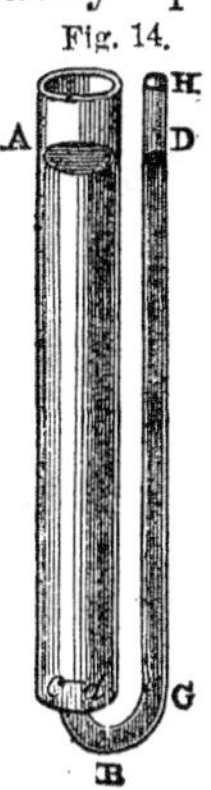

But suppose one arm of this tube to be larger than the other. Let the vessel have the form represented in Fig. 14, the arm A being twice as large as the other. To push the water down in H G, requires no greater effort with the breath than when the arms were of equal size. The downward pressure on one square inch at H, is transmitted as an equal upward pressure on *each* square inch at A, and thus a column twice as large is lifted one-half as high by the same force.

3. *The hydrostatic press.*—The hydrostatic press acts upon the principle just explained. It is a machine by which a small force may be made to exert a great pressure. Its construction may be understood by examining Fig. 15.

Two metallic cylinders, A and B, of different sizes, are joined together by a tube K. In the small cylinder there is a piston, *p*, which can be moved up and down by the handle M. In the large cylinder there is also a piston, P, having at its upper end a large iron plate which moves freely up and down in a strong framework, Q. Between the iron plate and the top of this framework, the body to be pressed is placed.

Now, when the small piston is raised, the cylinder A is filled with water drawn from the reservoir H, below, and when it is pushed down, this water is forced into the large cylinder, through the pipe K. There is a valve in this tube which prevents the water from returning, so that each stroke of the small piston pushes an additional quantity of water into the larger cylinder. By this means the large piston is pushed up against the body to be pressed.

To calculate the pressure exerted by the large piston, we must remember that the force acting upon the piston in A, will be exerted upon every equal amount of surface in B. To illustrate this: suppose the area of the large piston to be ten times the area of the small one; then one pound at A will produce a pressure of ten pounds at P. The handle, M, increases the advantage

Fig. 15.

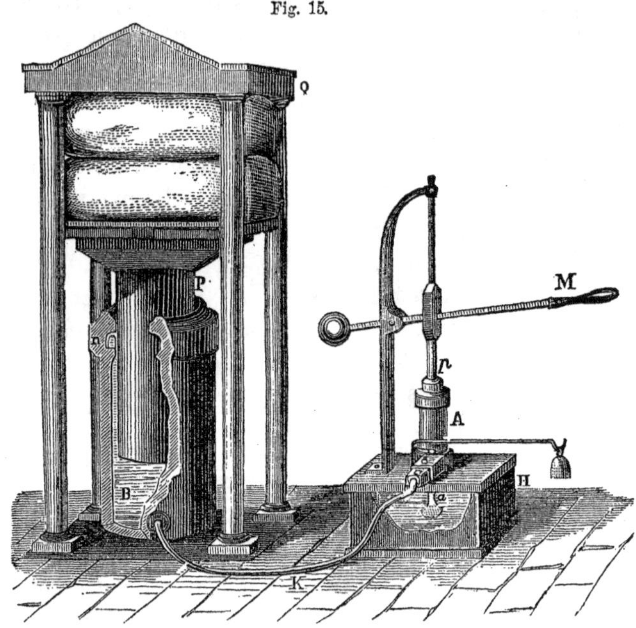

still more, according to the principle of the lever to be explained in a future chapter.

By increasing the size of the large cylinder, and diminishing the size of the small one, the pressure exerted by a given power will be increased proportionally. The weight of a man's hand might thus be made to lift a ship with all its cargo. The only limit to the increase of power would be the strength of the material of which the machine is made.

PROBLEMS ILLUSTRATING THE LAWS OF HYDROSTATICS.

1. A cylindrical vessel, whose base is 5 sq. ft., is 10 ft. high. It is filled with water. What pressure is exerted upon the base? *Ans.* 3,125 lbs.

2. If the bottom of a vessel has an area of 72 sq. in., and its top an area of 96 sq. in., and it is 9 in. high; what pressure will be exerted on the bottom when the vessel is full of water? *Ans.* 23.43 lbs.

3. Two vessels with equal bases are filled with water, one to a height of 9 in., the other to a height of 27 in. How many times more pressure on the base in the last case than in the first? *Ans.* 3.

4. How much pressure is being exerted against the *side* of a cubical vessel when full of water, its height being 18 inches? *Ans.* 105.46+ lbs.

5. How much pressure would be exerted upon 12 sq. in. of the sides of the vessel when the middle point of this surface is 20 inches below the top of the water? *Ans.* 8.68 lbs.

6. How many cubic inches of water will be displaced by a piece of pine wood weighing just 10 lbs? [See (13.)] *Ans.* 276.48.

7. How much less will a piece of marble measuring 100 cubic inches, weigh in water than in air? [See (13.)] *Ans.* 3.61 lbs.

8. The specific gravity of marble being 2.7, what will be the weight of 25 cubic ft.? *Ans.* 4,218.75 lbs.

9. How many cubic inches in a block of ice that weighs 75 lbs.? *Ans.* 2,229.12.

10. What is the specific gravity of flint glass if a fragment of it weigh, in air, 4,320 grs., and in water 3,023 grs.? *Ans.* 3.33.

11. The specific gravity of wax is to be found from the following data:—

Weight of the wax in air - - -	8 oz.
Weight of a piece of lead in air	16 oz.
Weight of the lead in water - -	14.6 oz.
Weight of wax and lead in water	13.712 oz.

Ans. 0.9.

12. A bottle holding 1,000 grs. of water is found to hold only 870 grs. of oil of turpentine. What is the specific gravity of this oil? *Ans.* 0.87.

13. How much pressure can be exerted upon the large piston of a hydrostatic press by applying 50 lbs. to the small piston; the area of the small piston being $\frac{1}{2}$ sq. in., that of the large piston 100 sq. in.? *Ans.* 10,000 lbs.

§ 4. OF THE PROPERTIES OF GASEOUS BODIES.

(16.) The most characteristic properties of gases are compressibility and expansibility. Besides these properties, gases possess others common to all forms of matter, among which we notice elasticity and weight.

1. *Compressibility.*—Let a small glass tube be fitted to the neck of a vial by a cork, so as to make an air-tight joint. Warm the vial *gently*, and then put a drop of ink into the top of the tube. As the vial cools the drop slowly moves down the tube until it finally stops at some point, A (Fig. 16). There it will be held by the capillary attraction of the tube. The air in the vial and tube up to the point A will thus be separated from the air outside. Now, closing the upper end of the tube with the lips, let the breath be gently pressed against the drop; it will be pushed down, it may be, a distance of several inches. The air in the vial can not escape, and the motion of the drop therefore shows that the air is being crowded into a smaller space, in other words, that it is *compressible*.

Fig. 16.

A

2. *Expansibility.*—If the vial (Fig. 16) be warmed by grasping it in the hand, or better, by standing it in warm water, the drop of ink will move upward in the tube. The air in the vial expands and pushes the drop along.

Or if, through the cork of a small bottle, a glass tube be passed, at the upper end of which is a bulb, and the lower end of which reaches down into the colored water contained in the bottle, the heat of a lamp flame may be applied to the bulb. It will be noticed that bubbles of air escape from the lower end of the tube. The air is *expanded*, so that the bulb and tube can no longer hold it all.

When the flame is withdrawn, the bulb gradually cools, and the water will rise in the tube and stand at a certain height, as shown in Fig. 17. Now, let the palm of the hand be laid upon the bulb; the water is driven down the tube by the expanding air. The gentle warmth of the hand is quite sufficient to produce a very considerable expansion of the air.

These two properties belong to solid and liquid bodies in various slight degrees, but pre-eminently to gases, which seem to be compressible and expansible *without limit.*

Fig. 17.

The force of heat in the last experiments is a repulsive force among the molecules of air, and pushes them farther apart. As long as this force increases by the action of the flame, it will push them farther and farther apart continually. In the first experiment, the slight pressure of the breath overcomes the repulsion among the molecules, and pushes them nearer together. Should the pressure be increased, we can give no reason why the molecules should not continue to approach each other. The limit of compressibility would be reached when the molecules should be brought into actual contact with each other; but to do this would doubtless require a pressure immensely greater than any at our command.

3. *Elasticity.*—The elasticity of air is beautifully shown by the simple apparatus already used (see Fig. 16) to illustrate the characteristic properties. When the breath is alternately pressed into and withdrawn from the tube, the air will alternately be compressed and spring back, the drop of ink jumping down and up in the tube to show it.

4. *Weight.*—The air has weight. If we would show it, we may first weigh an open vessel, properly ar-

ranged, and afterward take the air out of it and weigh it again, the difference in these two weights will be the weight of the air which the vessel contains.

But how can the air be taken out of a vessel? To answer this question we must become acquainted with the *air-pump*. A section of the essential parts of this important instrument is represented in Fig. 18. A cylinder, A B, is joined by means of a tube, *b e*, to a very smooth plate, *p*. A piston, *c*, moves air-tight in the cylinder. In the piston is a valve, *i*, which opens upward, and another valve at *b*, also opens upward from the tube into the cylinder. The vessel, *d*, from which the air is to be taken is placed upon the plate. Such vessels are usually called receivers.

Fig. 18.

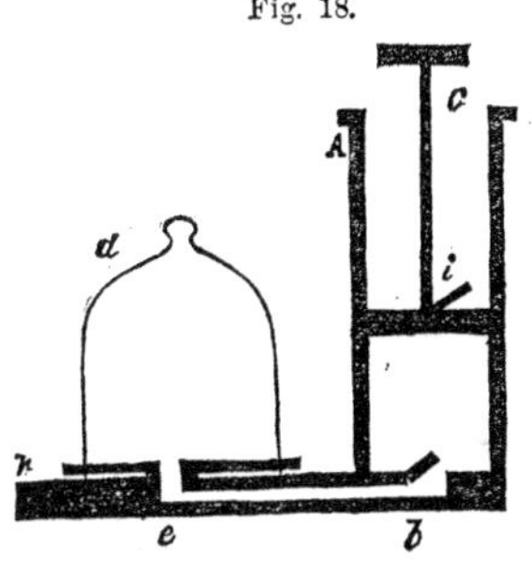

It will be seen that when the piston is raised, the valve, *i*, will be closed, and the air above it will be lifted out at the top of the cylinder. A vacuum would thus be formed below the piston were it not for the expansibility of the air in the receiver. This air expands, and a part of it is forced through the valve, *b*, into the cylinder. When the piston is pushed down, the air below it passes through the valve, *i*, and when by a second stroke the piston is lifted, this air is pushed out at the top of the cylinder, while another portion from the receiver is pressed through the tube into the cylinder below the piston. By each successive stroke, the quantity of air in the receiver is diminished, until, with a good instrument, the quantity left will be almost inappreciable. It is quite evident, however, that a *perfect* vacuum can not be obtained in this way. One form of this important instrument, complete, is represented in Fig. 19.

Fig. 19.

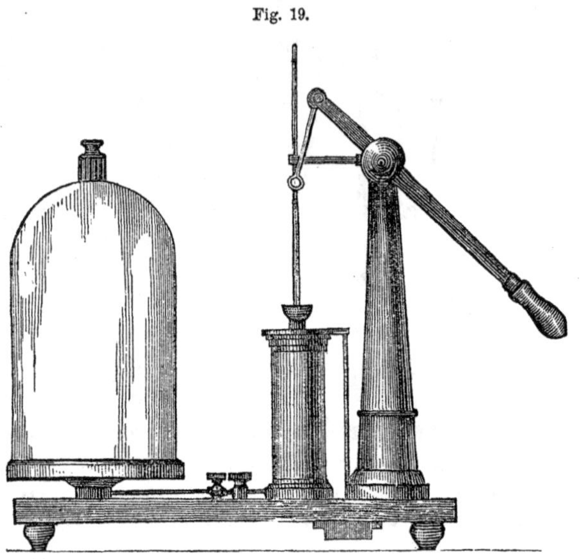

We may now attend to the process of weighing air. A hollow glass globe, with a stop-cock, is hung from one pan of a delicate balance, and its weight carefully found. It is then screwed to the opening in the plate of the air-pump, and the air is exhausted. The stop-cock is then closed to prevent the air from returning into the globe, which is then taken from the pump and weighed. It is found to weigh less than before, and the difference must be the weight of the air which has been taken out. At the ordinary temperatures of air, 100 cubic inches weigh about 31 grains.

§ 5. OF THE PRESSURE OF THE ATMOSPHERE.

(17.) The atmosphere exerts pressure in all directions. This pressure is about 15 lbs. upon every square inch of surface.

1. *The atmosphere exerts pressure.*—Since every one hundred cubic inches of air weigh about 31 grs., it is clear that the atmosphere must be exerting considerable pressure upon the surfaces of all bodies on which it rests.

This pressure may be shown in various ways. Take a glass tube of convenient length, open at both ends, and insert one end in a vessel of colored water. Apply the lips to the other end, and as the air is drawn out at the top, the water will rise rapidly in the tube. What pushes the water up? The ancients called it "Nature's abhorrence of a vacuum:" many at the present day are content to say that it is "sucked up." But let it be remembered that matter never moves unless it is *forced to move*, and that the forces of abhorrence and suction are simply fictions. The only force acting upon the water is the weight of the air resting on its surface in the vessel. This downward pressure pushes the water under the lower end and upward into the tube.

A more beautiful experiment consists in causing the pressure of the air to produce a fountain playing in a vacuum. A tall glass receiver (Fig. 20) is closed at the bottom by a stop-cock which terminates in a tube extending upward a little way into the receiver. The air from this receiver being taken by an air-pump, the stop-cock is immersed in a vessel of water and opened. Instantly the water leaps to the top of the receiver, and a beautiful fountain continues to play until the jet pipe is covered by the falling water.

Fig. 20.

2. *The pressure is in all directions.*—An experiment easily tried will show that the air is pressing equally in all directions. Stretch a piece of caoutchouc, or thin india rubber, over the large end of a lamp chimney, and firmly fasten it by winding a cord around it. Apply the mouth to the other end of the tube and draw the air out. The pressure of the air pushes the rubber into the tube. Hold the tube *in any position*, and in all positions the rubber will be pushed into the tube alike.

3. *The pressure is* 15 *lbs. to the square inch.*—If the air should be all taken out of our tubes used in the foregoing experiments [(17.) 1], the water would entirely fill them, and it is clear that the pressure of the atmosphere must, at least, equal the weight of the water in the tube. How much farther the water would rise if the tube was long enough, these experiments have not told. A *heavier* liquid will not be lifted as high as water, and will be more convenient for experiment. Mercury is a liquid metal about 13½ times as heavy as water, and it is found that the air will sustain a column about 30 inches high. The experiment is conducted as follows. Take a glass tube more than 30 inches long, closed at one end, and fill it with mercury. Close the open end with the finger and invert the tube. Now place the open end in a dish of mer-

cury and withdraw the finger. It will be seen that the top of the column of mercury in the tube, is about 30 inches above the surface of the mercury in the dish. (See Fig. 21.) The space above the mercury in the tube must be a vacuum.

Fig. 21.

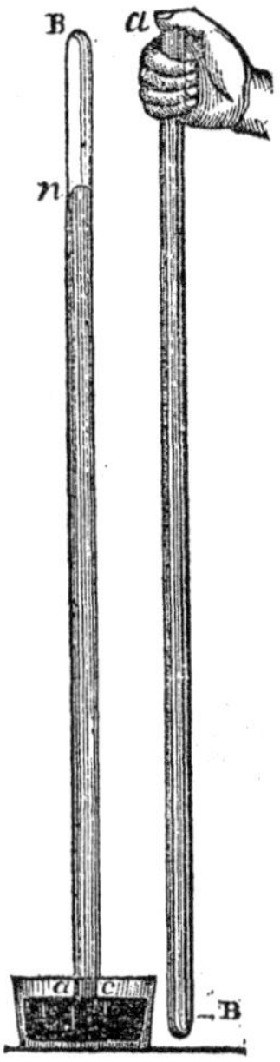

Now, the pressure of the atmosphere just balances the weight of this column of mercury. The weight of a column of mercury 30 inches high, the area of its base being one square inch, is 15 pounds. The downward pressure of the atmosphere is therefore 15 lbs. to the square inch of surface on which it rests.

(18.) The principle of atmospheric pressure is applied in the construction of many very useful instruments. We will notice the barometer, the common pump, the forcing pump, and the siphon.

I.—THE BAROMETER.

A.—The barometer column always indicates the pressure of the atmosphere. But the pressure of the atmosphere depends upon—

1st. Its weight.

2d. The amount of water-vapor in it.

3d. The elasticity of its lower portions, due to the action of heat.

1. *The barometer.*—If the apparatus used to deter-

mine the pressure of the atmosphere (see Fig. 21) is inclosed for protection in a frame of metal or wood, with a graduated scale attached, to measure the height of the column of mercury, it forms the instrument so well known as the *barometer*.

2. *Shows the pressure of the atmosphere.*—The pressure of the atmosphere is not always the same. When it is less than 15 lbs. to the inch, the column of mercury will be lower than 30 inches, and when greater, the column will be higher: indeed the height of the column will vary in exact proportion to every change in the pressure of the air which supports it. But notice that when the mercury sinks in the tube, it must rise in the cistern, and that, hence, the column must shorten at *both ends*, while the figures on the scale only show the change which takes place at the top: they fail to tell the true height of the column.

This error is avoided in what is called Fortin's barometer, by means of a cistern with a flexible bottom (see Fig. 22). The bottom of this cistern is made of deerskin, and rests upon the end of the screw C, by which it may be lowered or lifted. An ivory pointer, A, is fastened to the top of the cistern, and its lower end is the point from which the distances are measured on the scale which shows the height of mercury in the tube. If the surface of the liquid in the cistern just touches this point, then the figures on the scale show the true height of the column, which indicates the pressure of the atmosphere.

Fig. 22.

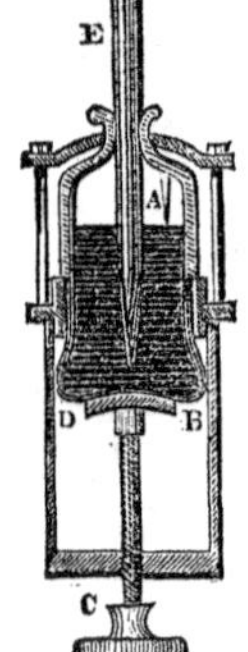

3. *The pressure of the atmosphere depends upon its weight.*—Consider the atmosphere as a vast ocean of

air, whose depth (or height, since we are at the bottom of it) is thought to be about fifty miles. Its upper surface can no more be at rest than can the surface of the sea, and its billows must be more immense, because its substance is more easily moved. The barometer over which these great waves sweep to and fro, being now under the crest, and then under the depression, is subject to the pressure of columns of air of different heights, and the mercury must rise or fall accordingly.

Again; the weight of the atmosphere will vary with the altitude of the place where the observation is made. When we go up a mountain-side, we leave a part of the atmosphere below us, and, of course, the height of the column above us is less. The barometer column will, therefore, be shorter.

Upon this principle the barometer is *used to measure the height of mountains.* If the density of the atmosphere were uniform, the fall of the mercury would be in exact ratio of the distances upward, and knowing the height required to make the mercury fall $\frac{1}{10}$ of an inch, this multiplied by the number of tenths through which it is observed to sink, would tell the height of the mountain. The truth is, however, that the density of the air rapidly diminishes as we ascend. Temperature, too, affects its pressure. In spite of these difficulties, tables have been constructed, by which the height of a place above the sea-level may be calculated, by observing the height of the barometer column and the temperature of the atmosphere. (See Cooke's Chem. Phys., p. 511.)

4. *Upon the amount of water-vapor it contains.*—Mixed with the air, at all times, are considerable quantities of *invisible* vapor of water. If the atmosphere was pure dry air alone, it would exert a certain press-

ure: if it consisted wholly of water-vapor, it would exert a different amount of pressure: it does consist of a mixture of these two gases, and the pressure it exerts is the *sum* of the pressures they would separately exert.

It follows that the atmospheric pressure will be greatest when there is the greatest quantity of water-vapor in the air; the barometer column will then rise. But let this vapor be condensed into clouds, and it will have but little force of elasticity, and will exert but a small fraction of its former pressure; hence the barometer column will stand lower in cloudy weather.

On this principle the barometer is used to indicate changes in the weather. *A rising column indicates fair weather; a falling column indicates foul weather.*

Fig. 23.

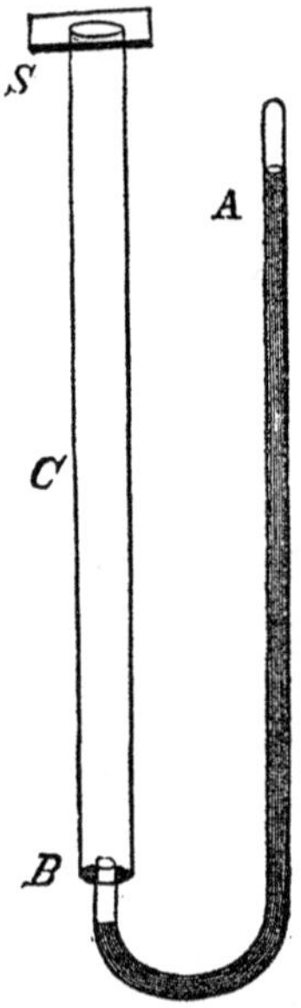

S

A

C

B

This rule is to a great extent reliable. Others are given by different observers, but they must be taken with considerable allowance.

5. *Upon the elasticity of its lower portions.*—Let us approach this topic by means of an experiment. The tube, A B, Fig. 23, having been filled with mercury and inverted, the space above A is a vacuum, and the column of mercury is sustained by the pressure of the air at B. Let this end pass, air-tight, through the stopper in the lower end of a long glass tube C. The upper end of this tube is ground smooth, and covered with a heavy ground glass slide S. Now let the tube

C, be heated. The air within is expanded; it can not escape at S; the entire expansive force is exerted upon the mercury at B, and the column shows this pressure by rising at A. While the heat is continued, let the slide S be drawn so as to leave a very small hole in the top of the tube C; a gradual fall of the mercury at A will show that the pressure of the air is diminishing.

Now the atmosphere is heated by coming in contact with the earth: the lower stratum is heated *first*, and the upper strata in succession afterward. The heated stratum is expanded like the air in the tube C, when heated. It attempts to rise, but the strata above, *not yet heated*, rest upon it and prevent its rising, just as the slide S keeps the air in our tube, and the expansive force of this stratum must raise the barometer column. This expansive force will, for a time, increase, until it is strong enough to lift the weight of the upper strata of air, after which it will diminish, just as it did when the slide S was removed.

When the atmosphere *cools*, the lower stratum is first *condensed*, and this allows the air above to move downward. In falling it gains a certain velocity, and exerts a greater pressure: the barometer column must be raised thereby.

We see, then, that as the atmosphere is daily warmed and cooled, the barometer column must rise and fall. Observation confirms this. From nine o'clock to ten in the morning, the mercury reaches its greatest height; afterward it begins to fall, and at three or four in the afternoon it reaches its lowest point. It then begins to rise again, and reaches its greatest height at nine or ten in the evening. These motions occur every day, and are so regular that, as Humboldt says, they might be used to indicate the time of day, only that the distance through which the column fluctuates is very small, being greatest at the equator, where it amounts to $\frac{12}{100}$ of an inch.

II.—THE COMMON PUMP.

This instrument, as generally made, consists of two cylinders or barrels, A and B, Fig. 24, with a valve, S, at their junction, opening upward. In the upper barrel is a piston, P, in which is a valve, O, also opening upward. The piston is moved by means of the handle H, and the water may flow from the spout C.

Fig. 24.

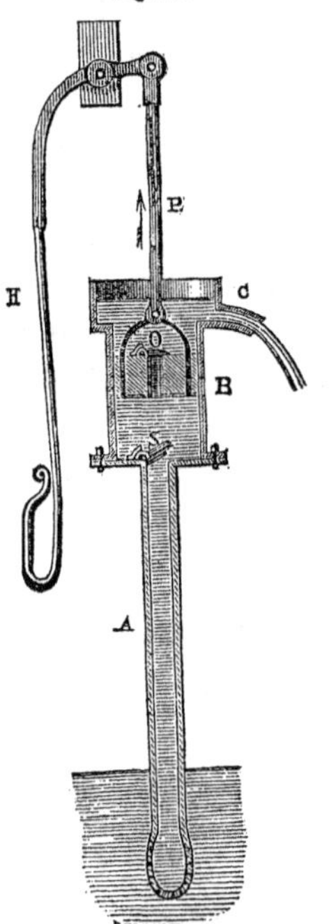

When the piston is lifted, the air above it will be lifted out of the barrel. A partial vacuum will thus be formed below the piston, and the pressure of the air upon the surface of the water in the well, will push the water up the barrel A, through the valve S, into the barrel B. When the piston goes down, the valve S will close, and prevent the return of the water to the well. The valve in the piston will be opened, and the water will pass through it. When the piston is again lifted, the water, now above it, will be lifted to the spout, while the atmospheric pressure will force another portion into the barrel below the piston.

At the sea-level the pressure of the air will sustain a column of mercury about 30 inches high. Since mercury is, at ordinary temperature, about $13\frac{1}{2}$ times heavier than water, the same force will lift a column of water $13\frac{1}{2}$ times as high: $13\frac{1}{2} \times 30 = 405$; 405 in. $= 33\frac{3}{4}$ft. The lower barrel of the common pump must not exceed $33\frac{3}{4}$ft. in length, even at the level of the sea.

III.—THE FORCING PUMP.

In the forcing pump the piston has no valve, but from near the bottom of the upper barrel there is a tube passing to an air-chamber, with a valve opening into the chamber. A section of this instrument is represented in Fig. 25. Reaching from near the bottom of the air-chamber I K, is a tube, L M, which extends to any place at which the water is to be delivered.

Fig. 25.

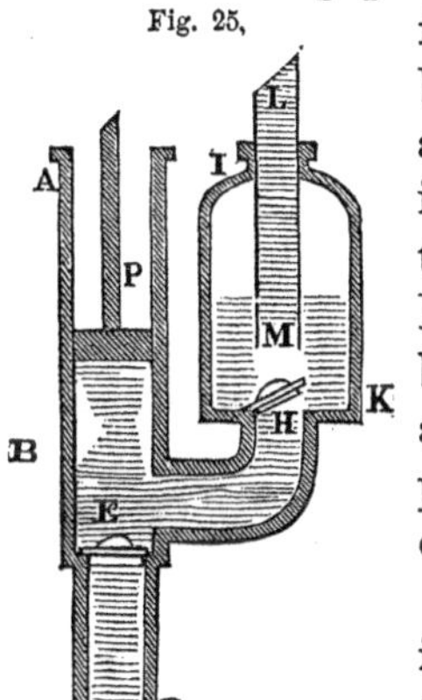

Now, when the solid piston P, is raised, water is pressed through the valve E, into the barrel B. When the piston is pushed down again, the water is driven through the tube into the air-chamber, and compresses the air in it. By every stroke, the water accumulates in the chamber, and the air is more and more compressed. The pressure of this condensed air upon the water in the chamber, pushes it up through the tube L M, to the place where it is desired. Without the air-chamber, the water would issue from the pipe in jets; with the chamber, the water issues in a steady stream.

IV.—THE SIPHON.

The siphon is an instrument by which liquids may be transferred from one vessel to another, by atmospheric pressure. It consists of a bent tube, one arm of which is longer than the other. In Fig. 26, the siphon in operation is shown. Having been first filled with water, its short arm is inserted in the water to be transferred from the vessel, C, and it is then found that the water will flow steadily, until the lower end of the short arm is left uncovered, or, in other cases, until the water in the two vessels stands at the same level.

Fig. 26.

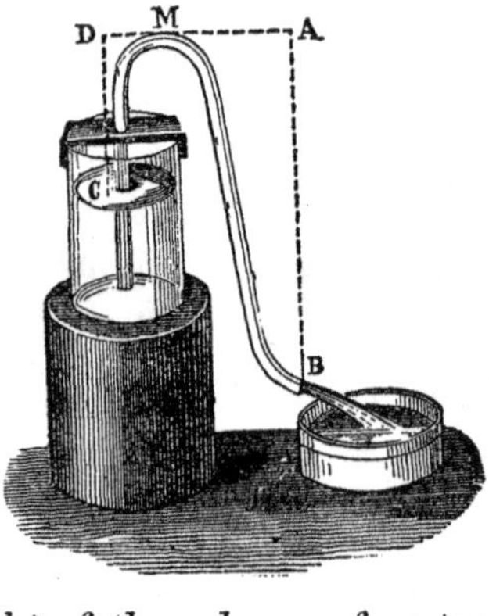

Now, the downward pressure of the air at C, is partly balanced by the weight of the column of water in the short arm of the tube; the *excess* of force will tend to push the water over through the bend. On the other hand, the atmospheric pressure at B is partly balanced by the weight of the water in the long arm; the *excess* will tend to push the water back through the bend toward C. It is clear that the pressure of air, minus the weight of the *shorter* column of water, is more than the same pressure, less the weight of the *longer* column, and hence that a greater force will be exerted to push the water from C toward B, than from B toward C: the liquid will flow in the direction of the greater force.

§ 6. OF THE RELATION BETWEEN THE VOLUME AND THE WEIGHT OF AIR.

(19.) The volume of any given weight of air, or other gaseous body, will vary with every change in the pressure or the temperature to which it is subjected.

I.—PRESSURE.

A.—The volume of a given weight of air will be inversely as the pressure upon it. Hence the density of the atmosphere is greatest at the surface of the earth.

1. *Volume inversely as the pressure.*—Press the breath into the tube above the drop of ink (see Fig. 16), and the air in the bottle will be *condensed.* Now draw the air out of the tube, and the drop rises, showing that the air below has *expanded.* The same quantity of air is here seen to fill *less* space, when the pressure upon it is increased, and *more* space when the pressure is diminished.

Now, we may prove by experiment, first, that with a double pressure, the volume will be just one-half; and, second, that with half the pressure, the volume will be just double.

In the first case, we use a bent glass tube (Fig. 27),

the short arm being closed, and the other, which should be more than 30 inches long, being open at the top. A graduated scale, to which the tube is firmly bound, measures inches from the bend.

Fig. 27.

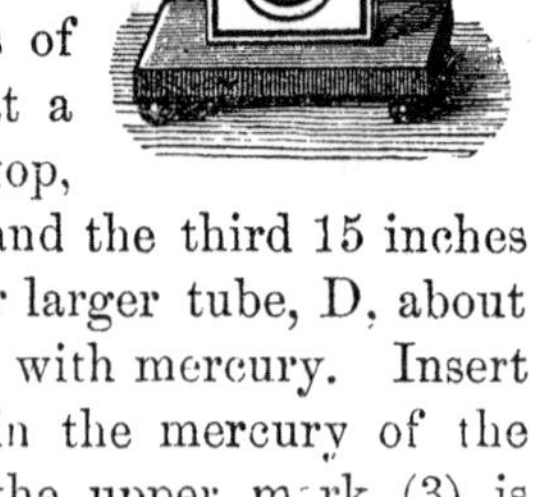

Now, let mercury be poured into the tube, until it fills the bend. The air presses upon the mercury in the long arm, and this liquid transmits the same pressure to the air in the short arm. The pressure upon the air in the short arm is, therefore, 15 lbs. to the square inch. If we fill the long arm, as shown in the figure, to the height of 30 inches, with mercury, we will be adding a pressure of 15 lbs. to the inch. The pressure upon the air in the short arm, will then be *doubled*, and we discover that the mercury has risen, crowding the air before it, and stands at A, the air filling just *half* the original volume.

In the second case, we take a glass tube, A B (Fig. 28), about 25 inches long, and open at both ends. Let three narrow bands of paper be pasted upon it, one at a distance of 3 inches from the top, another 6 inches from the top, and the third 15 inches from the second. Let another larger tube, D, about 30 inches long, be nearly filled with mercury. Insert the end A, of the small tube in the mercury of the other, and push it down until the upper mark (3) is at the level of the mercury. Now, clasping the finger tightly over the end B, thus inclosing 3 inches of air in the tube, lift it until the third mark is brought up to the top of the mercury. The air will be found to fill the space of 6 inches.

Fig 28.

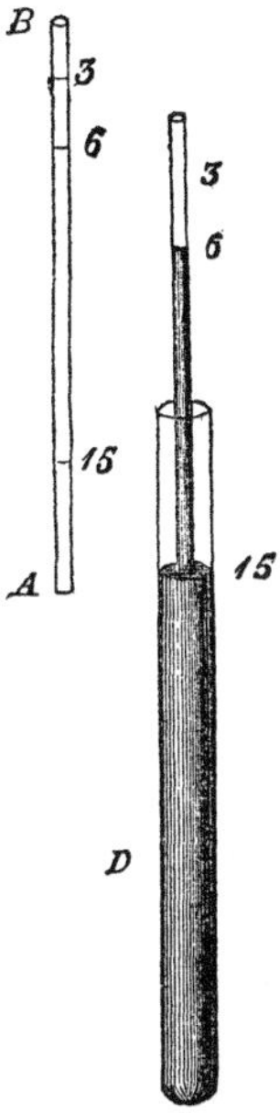

B 3 6 15 A 3 6 15 D

Before the tube was lifted, the pressure of the atmosphere, 15 lbs. to the inch, was exerted upon the air in it: after the tube is lifted, the atmosphere sustains a column of mercury 15 inches high. To do this takes half the pressure it can exert, the other half is exerted upon the air above the mercury. We thus show that with *half* the pressure the volume will be just *double.*

From these two experiments we infer that *the volume of a given weight of air will be inversely as the pressure upon it*, and repeated experiment confirms the inference.

This law was discovered by the Abbé Mariotte in France, and is generally called Mariotte's law. (See Cooke's Chem. Phys., p. 287.)

2. *The density of the atmosphere.*—When a given weight of air is crowded into one-half its original volume, it must be twice as dense; and when expanded into double its first volume, it can only be half as dense. The density of air will therefore be exactly in proportion to the pressure upon it. So the atmosphere, where its pressure is greatest, will be most dense.

3. *Is greatest at the surface of the earth.*—The atmosphere in contact with the earth is pressed upon by

all the air above, even to the top of the atmosphere. At a distance above the earth, the atmosphere receives less pressure, because there is less air above to exert it. The density being greatest where the pressure is greatest, the air at the surface must be more dense than the portions above. The air is much less dense at the top of a high mountain than at its base.

II.—TEMPERATURE.

B.—The volume of a given weight of air will be greater as its temperature is higher. It expands $\frac{1}{490}$ of its bulk for every additional degree of heat.

1. *Heat increases the volume of air.*—Let the palm of the hand be laid upon the bulb (Fig. 17), and the fluid in the tube descends, because the air in the bulb expands. Pour cold water upon the bulb, and the fluid ascends because the air above it is condensed. Apply the heat of the lamp flame to the bulb, and the water in the tube will be quite driven out at the bottom: let it cool again, and the water rises to its former height. These experiments show that the addition of heat expands air, and that its withdrawal contracts it.

2. *At the rate of $\frac{1}{490}$ its bulk for each degree.*—The expansion of air and other gases by heat, is uniform. One degree of heat, when the temperature is low, produces the same expansion as one degree, when the temperature is high. If we have 490 cubic inches at a temperature of 32°, it will become 491 cubic inches if heated one degree, making its temperature 33°. In other words, it expands $\frac{1}{490}$ of its bulk *at* 32°, for each additional degree of heat applied.

PROBLEMS ILLUSTRATING THE LAWS OF GASEOUS BODIES.

1. What is the weight of a cubic foot of air at ordinary temperature and pressure?

Ans. 535.68 grs.

2. What is the weight of 100 cub. in. of oxygen gas at ordinary temperature and pressure, its specific gravity being 1.108? *Ans.* 34.348 grs.

3. What is the weight of 100 cub. in. of nitrogen gas at ordinary temperature and pressure, its specific gravity being .692?

4. What pressure will be exerted by the atmosphere on a surface of 1 sq. ft.? [See (17.)] *Ans.* 2,160 lbs.

5. What pressure does the atmosphere exert upon a square inch surface when the barometer column is 28 inches high? [See (18.) A. 2.] *Ans.* 14 lbs.

6. How high a column of water would the atmosphere sustain when the barometer column stands at a height of 28 inches? *Ans.* 31½ ft.

7. Suppose 100 cub. in. of air at a pressure of 15 lbs. to the inch is made to receive an additional pressure of 15 lbs. to the inch, what will be its volume? [See (19.) A.] *Ans.* 50 cub. in.

8. How much pressure must be removed from 100 cub. in. of air, at usual density, in order that it may expand to a volume of 200 cub. in.?

Ans. 7½ lbs. to the inch.

9. In the air-chamber of the forcing-pump, the air is compressed into half its former bulk; how high will the water be thrown? *Ans.* 33¾ ft.

10. If we have 500 cub. in. of air at 32° temp., how much will there be when it is heated to a temperature of 75°? [See (19.) B.] *Ans.* 543.88+ cub. in.

PART II.

THE PHENOMENA OF BODIES IN MOTION.

CHAPTER III.

OF MOTION.

INTRODUCTION.—APPLICATION OF THE FUNDAMENTAL IDEAS.

(20.) Read (4.) and (7.)—Attraction and repulsion acting upon masses of matter determine their condition of rest or motion.

1. The motion of bodies falling to the ground is due to the *attraction* of gravitation. The motion of air in wind is caused chiefly by the *repulsive* power of heat. The bullet speeds on its mission of death, urged by the *repulsive* force of exploding gunpowder. The forces which produce the endless variety of motions in nature are found, when carefully studied, to be only different forms of attraction and repulsion.

We speak of the *forces of nature*, and call them wind, water, gravitation. This is well, because these names have been given to familiar forms of force. We will continue to use these terms: at the same time, let us do justice to the simplicity of God's stupendous works, by remembering that the forces of nature are *attraction* and *repulsion*.

§ 1. OF MOTION CAUSED BY A SINGLE FORCE.

(21.) There are three important principles called the laws of motion :—

1st. A body at rest will remain at rest; or, if in motion, it will move forever in a straight line, unless acted upon by some force to change its condition.

2d. A given force will produce the same amount of motion, whether it act upon a body at rest, or in motion.

3d. Action and reaction are equal, and in opposite directions.

1. *The first law.*—The first law is proved when we remember, that the inertia of matter forbids that a body shall, in any way, change its own condition.

Then, why are bodies so constantly changing their condition of rest or motion? Who ever saw a body in nature, moving in an absolutely straight line? Bodies are constantly under the influence of forces which do change their condition. A stone thrown from the hand would move forever in a straight line, if it felt only the force of the hand; but gravitation, and the resistance of the air, compel it to move in a graceful curve instead. The pleasing variety of natural motions is brought about by the unceasing action of external forces.

2. *The second law.*—Let us examine this law by means of the diagram, Fig. 29. If a ball be thrown suddenly from the point A, horizontally, it would go to the point B, if it could be let alone by other forces. So likewise if it be dropped from A, gravitation alone will carry it to C. Now, these separate effects of the two forces will be exactly produced when the forces act together. Suppose the ball to go from A to B in one

minute, and, when dropped, to fall from A to C in the same time. Now, while the ball is moving toward B, gravitation is pulling it downward, and at the end of the minute it will be found at D, having moved to the right, a distance exactly equal to A B, and downward

Fig. 29.

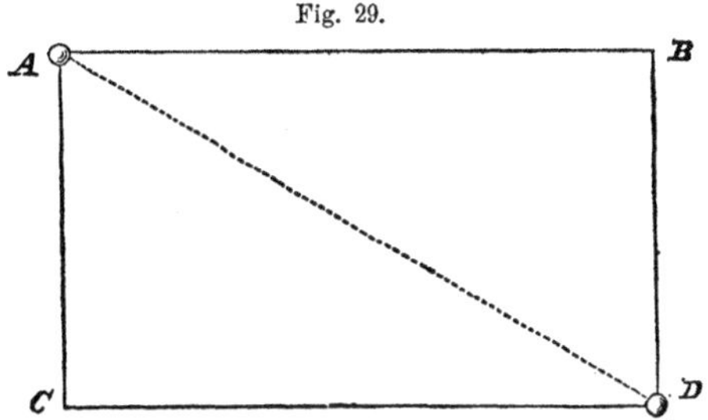

through a distance exactly equal to A C; so that the force of gravitation produces the same effect, whether it act upon the ball *resting* at A or *in motion* toward B.

3. *The third law.*—If a table be struck, the hand that strikes it receives a blow as well. The hand *acts* upon the table; the table *reacts* upon the hand. Attend, now, to the following experiment. Two ivory balls are suspended by cords, and hang in contact against a graduated arc. When the ball, B, is lifted up the arc to D, and then allowed to swing against the other, it strikes it and instantly stops, while the other ball takes up its motion, and goes to the point C. The first ball acts upon the

Fig. 30.

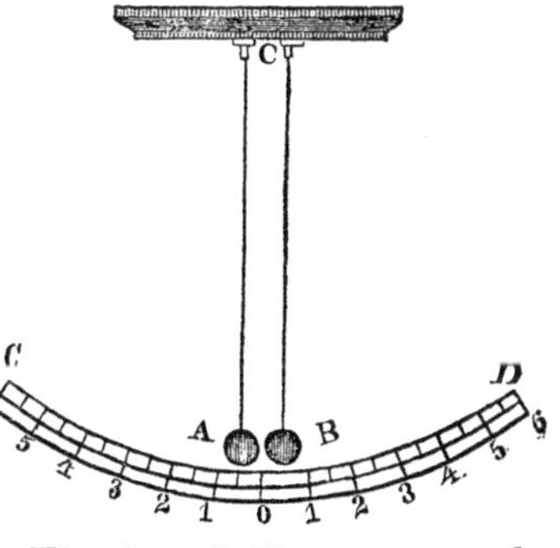

second; the second reacts upon the first. Now, if we notice that the motion from D to B, which is *stopped* by the reaction of the second ball, is just equal to the motion from A to C, which is *caused* by the action of the first, it becomes evident that the two forces must be equal, and exerted in opposite directions.

It follows from this principle that, when two bodies come in contact, each one gives and receives an equal shock. The hand which strikes the table is itself bruised, and the bullet which shatters the bone, is itself battered and torn.

(22.) The velocity of a moving body will be uniform if it be produced by an impulsive force and opposed by no resistances.

The elements of motion are *time*, *space*, and *velocity*. In uniform motion, the space is equal to the product of time multiplied by velocity.

1. *Velocity.*—Velocity, in a popular sense, is simply rapidity of motion, but if the term is to be of any scientific value it must be more definitely applied. Velocity is the *distance* passed over by a body in a *unit of time.* The velocity of a cannon ball, for example, may be 2,000 ft. a second: that of a train of cars may be 30 miles an hour.

2. *Uniform velocity.*—In uniform velocity, a body moves over equal spaces in equal times. If, for instance, in each of three successive hours, a steamboat travels 15 miles, its velocity is uniform.

3. *An impulsive force.*—An impulsive force is one which, after acting for a time, ceases. The stroke of a bat, which knocks the ball, is an impulsive force; so are the blows of a hammer. No matter how

long a force may have been acting, if it be suddenly withdrawn, it is at that moment an impulsive force.

4. *Uniform motion produced by an impulse.*—If a body can be free from all forces but the impulse which gives it motion, its velocity will be uniform. This seldom occurs. How rarely do we see a uniform motion produced by an impulse, either in nature or in art! It is because all bodies are under the influence of several forces at once, such as gravitation, friction, and the resistance of air, by which their velocities are changed. The motion of the earth on its axis is, however, a sublime example of uniform motion.

In the arts, a uniform motion can be secured only by the constant application of power. The impulse which starts a train of cars, would make it move uniformly if it did not meet with resistances: to overcome these, a constant pressure of the steam must be applied. If this pressure be, at all times, just equal to the purpose, the motion of the train will be uniform.

5. *Space equals time multiplied by velocity.*—It is evident that a train of cars, going uniformly at the rate of 25 miles an hour, will, in ten hours, go 250 miles: $250 = 25 \times 10$, or the space is equal to the product of the two other elements, time and velocity.

We may express this principle by the simple equation—

$S = T \times V$: in which
S stands for Space.
T " " Time.
V " " Velocity.

Now, if any two of these elements are given, the third may be found by *substituting given values for the letters, and then performing the operations indicated.* For example, what is the velocity of a bullet which goes 2,000 ft. in 20 seconds, supposing its velocity uniform? The value of S is 2,000 ft.; the value of T is 20 seconds. Putting these values in the equation, it becomes

$$2{,}000 = 20 \times V. \quad \text{Hence } V = 100.$$

(23.) The motion of a body produced by the action of a constant force alone, will be uniformly accelerated. The difficulties in the way of any accurate experiment upon uniformly accelerated motion are overcome by Atwood's machine.

1. *A constant force.*—By a constant force we mean a force which acts upon a moving body all the time alike. The force of gravitation is the most perfect example of a constant force.

2. *Uniformly accelerated motion.*—The motion of a body is uniformly accelerated, when its velocity increases equally in successive units of time; as, for example: 5 ft. the first second, 8 ft. the next second, 11 ft. the third second, 14 ft. the fourth, and so on.

The motion of a falling body is the most perfect example: of uniformly accelerated motion. It would be a perfect example were it not for the resistance of the air.

3. *Difficulties in the way of experiment.*—Three difficulties are in the way of accurate experiments upon the motion of a body falling. 1st. It is so rapid that no accurate observations can be made. 2d. It is subject to the resistance of air, which reduces its velocity. 3d. The friction of any apparatus used is likely to impede it.

4. *These difficulties overcome by Atwood's machine.*—These difficulties are, for the most part, overcome by Atwood's machine (Fig. 31). Two heavy weights A and B, are fastened to the ends of a small cord which passes over a grooved wheel, D. Each end of the axis of this wheel rests on the circumferences of two other wheels. The standard L, is graduated: upon it is a movable ring, C, which allows the weight A to pass through it, and a table below, which arrests the motion of the weight at any desired point. The time of motion is measured by the pendulum F.

Fig. 31.

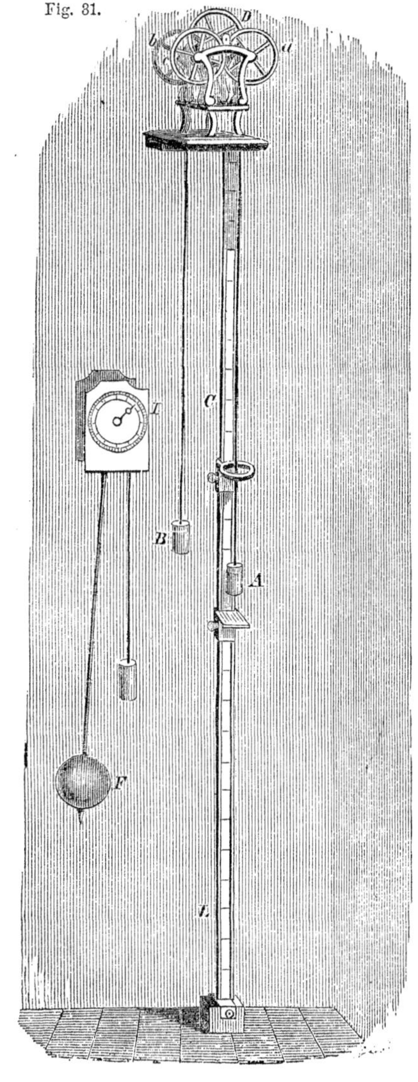

The two weights A and B are made exactly equal, and, of course, when left to themselves, will remain at rest. But if a small bar of brass be laid upon the weight A, motion takes place, due entirely to the action of gravitation upon the bar.

Now, suppose the large weights each to be $31\frac{1}{2}$ oz., and the weight of the small bar to be 1 oz. When they all move, 64 oz. are in motion, caused by the force which acts upon the 1 oz. bar. It is evident that 64 oz. will move only $\frac{1}{64}$ as fast as 1 oz., with the same force. The motion of the weights is produced by a *constant force*, gravitation; but it is only $\frac{1}{64}$ as rapid as when the bodies fall freely. A *slow* motion is thus obtained. The resistance of the air against the small surfaces of the ends of the heavy weights, is very slight when they move slowly; the friction of the wheels at the top is trifling, and thus the three difficulties in the way of experiment are overcome.

(24.) By experiments with Atwood's machine we may prove:—

1st. That a body moving under the influence of gravitation during any interval of time, will gain a velocity which, acting alone, will carry the body twice as far in the next equal interval.

2d. That gravitation will add to the motion of a body just as much in every interval of time as it produced in the first.

By the help of these principles we may analyze the motion of a falling body. From the diagram which represents this analysis, we may construct a table which shall contain the values of time, space, and velocity; and from this table obtain the laws which govern the motion, and the formulas by which problems may be solved. (See Cooke's Chem. Phys., p. 23.)

1. *Proof of the first principle.*—Let the weight A, carrying the small bar, be brought to the top of the graduated standard. Suppose that, in one second after its release, it falls to the ring C, a distance of 3 inches. The small bar will be caught off by the ring; the weight A will pass through, and in the next second it will be found to go exactly 6 inches. By putting the ring at different places on the standard, it will be found that, in every case, as in the one just described, the *body moving under the influence of gravitation during any interval of time, will gain a velocity which, alone, will carry the body twice as far in the next equal interval.*

2. *Proof of the second principle.*—If the weight and bar fall 3 inches in one second, they will be found to fall 12 inches in two seconds. The distance fallen in the *second interval* is 9 inches. If the bar were taken off at the end of the first second, the weight would go alone, 6 inches in the next. It is clear, then, that the bar acting in the last second adds a motion of 3 inches, the same amount as it produced in the first. Repeated experiments show that *gravitation will add to the motion of a falling body just as much in each second as it produced in the first.*

3. *Analysis of the motion of a falling body.*—Now suppose a body to fall from the point A (Fig. 32), toward the point D. In the first second it will fall a certain distance, which we will represent by A B. For a moment suppose the force of gravity should cease to act, the body would still move on, and we know (by the first principle), that it would go in the next

second, just twice as far as it did in the first. Then mark below B, two spaces, each equal to A B, to represent this distance, and mark it with a heavy line, that the eye may see at a glance, that it is the distance due to velocity alone. But we know (by the second principle) that gravitation in this second will add a space just equal to A B. Marking this space in the figure, we find that in two seconds the body will fall to C.

Fig. 32.

A

B

C

D

In *two seconds* the body has fallen 4 spaces, in the *next two seconds* it will go twice as far, 8 spaces, by velocity alone. In the *first* of these two seconds, which is the *third second* of its fall, the body will go 4 spaces by velocity. The force of gravity adds another space, so that at the end of 3 seconds the body will be found at D.

To find the distance passed in the 4th second, notice that in the first 3 *seconds* it has passed 9 spaces; that in the next 3 *seconds* it will go, by its velocity alone, 18 spaces, and that in *one* of these 3 seconds, which would be the 4th second, it would go 6 spaces. Mark 6 spaces for velocity, and add one for the action of gravitation.

4. *Construction of the table.*—Now, in this diagram the values of time, space and velocity stand clearly before us, and we may put them in a tabular form. In the first column, headed T, put the number of seconds, 1, 2, 3, 4. In the second column,

T.	S.	V.	*s.*
1	$1g$	$2g$	$1g$
2	$4g$	$4g$	$3g$
3	$9g$	$6g$	$5g$
4	$16g$	$8g$	$7g$

headed S, put the total space passed over at the end of these seconds, representing the distance A B by g. In the third colum, headed V, put the velocities gained at the end of each of the seconds. And, finally, in the fourth column, headed s, put the spaces passed in each separate second.

5. *From the table obtain the laws.*—The relation between time, space, and velocity, may be seen by comparing their values given in this table.

Notice, first, that the values of S have the same ratio as the squares of the value of T. For instance, take the spaces $4g$ and $9g$ with the corresponding times 2 and 3, we find that $4g : 9g :: 2^2 : 3^2$. Hence, *the spaces passed by a falling body in different times are as the squares of the times.*

Notice, second, that the values of V have the same ratio as the values of T. Thus we find that the velocities, $4g$ and $6g$, have the same ratio as 2 and 3, the corresponding values of time. Hence, *the velocities of a falling body at the end of successive intervals of time will vary as the time of fall.*

Notice, third, that *the spaces passed in separate seconds* (the values of s) *are as the odd numbers*, 1, 3, 5, 7, &c.

6. *From the table also obtain the formulas.*—It will be seen that by squaring any one of the values of T in the table, and then multiplying by g, the corresponding value of S will be obtained. Hence,

$$S = T^2g \ (1).$$

Again, we may discover that if the value of T in any case be multiplied by $2g$ the corresponding value of V will be produced. Hence,

$$V = 2gT \ (2).$$

We see, again, that if the value of S in any case be multiplied by g, the square root of this product multiplied by 2, gives the value of V. Hence,

$$V = 2\sqrt{Sg} \ (3).$$

Finally, a little attention will show, that if the value of T in any case be multiplied by 2, the product diminished by 1, and the remainder multiplied by g, the corresponding value of s will be obtained. Hence,

$$s = (2T - 1)\,g \ (4).$$

7. *By these formulas solve problems.*—By the use of these four formulas all problems in uniformly accelerated motion may be solved. In all cases, g represents the distance passed by the body in the *first* interval of time. Its value will be different for different forces. When gravitation is the constant force which causes the motion, the value of g is $16\frac{1}{12}$ feet.

A single illustration will show how the formulas may be used. If a stone be dropped into a well whose mouth is $144\frac{3}{4}$ ft. above the water, how long will it take to reach the water? Since gravitation produces this motion, the value of g is $16\frac{1}{12}$ ft. The $144\frac{3}{4}$ ft. is the value of S, and the value of T is required. The relation between these elements is expressed by the formula $S = T^2g$, and by substituting the given values we have $144\frac{3}{4} = T^2 \times 16\frac{1}{12}$. The value of T, from this equation is 3 seconds.

PROBLEMS ILLUSTRATING THE LAWS OF MOTION WHEN PRODUCED BY A SINGLE FORCE.

1. A body moves uniformly over a distance of 780 feet with a velocity or 5 feet a second: in what time did it go? *Ans.* 156 sec.

2. Under the influence of an impulsive force, a body moves at the rate of 25 feet a second: how far will it go in one minute?

3. A stone dropped from the top of a tower, struck the ground in 4 seconds: how high is the tower?

Ans. $257\frac{1}{3}$ feet.

4. If the tower were $257\frac{1}{3}$ feet high, with what velocity would a stone strike the ground?

Ans. $128\frac{2}{3}$ feet.

5. If the velocity of the stone should be $128\frac{2}{3}$ feet a second; how long a time had it been falling?

Ans. 4 sec.

6. A body falls 4 seconds: how far does it go in the fourth second? *Ans.* $112\frac{7}{12}$ feet.

7. Under the influence of a constant force, a body moves 3 feet the first second: how far will it go in 5 seconds? *Ans.* 75 feet.

8. A body is falling toward the earth; it is at the same time moving horizontally under the influence of a constant force which made it go 10 feet in the first second: how far, horizontally, did it go in 8 seconds? [See (21.) 2.] *Ans.* 640 feet.

9. How far did it fall in the same time?

Ans. $1{,}029\frac{1}{3}$ feet.

10. With what velocity did it strike the ground?

Ans. $257\frac{1}{3}$ feet.

11. What velocity did it gain in a horizontal direction? *Ans.* 160 feet.

12. How far did it go horizontally in the 5th second?

Ans. 90 feet.

13. How far did it fall in the 5th second?

Ans. $144\frac{3}{4}$ feet.

14. Under the influence of a constant force, a body

goes 12 feet in the first 3 seconds; how far does it go in 18 seconds? *Ans.* 432 ft.

15. A ball is thrown directly upward, starting with a velocity of 96½ feet, to what height will it rise? *Ans.* 144¾ ft.

The motion of this ball thrown upward, will be *retarded* by gravitation, in exactly the same ratio that it is *accelerated* in falling to the ground again. The height to which it rises is the same as that from which it falls. This problem may be solved exactly as if the question were: from what height would the ball fall to gain a velocity of 96½ feet a second?

16. A ball is shot upward with a velocity of 386 feet: how long will it continue to rise? *Ans.* 12 sec.

17. How high does it go? *Ans.* 2,316 ft.

18. How long does it remain in the air? *Ans.* 24 sec.

19. How far does it rise in the *last* second of its ascent? *Ans.* $16\frac{1}{12}$ ft.

20. How far does it fall in the *last* second of its descent? *Ans.* $369\frac{11}{12}$ ft.

21. Suppose the large weights of Atwood's machine to be each 31½ oz., and the weight of the small bar to be 1 oz. We find, by experiment, that the weight and bar go 3 inches in the first second: what is the force of gravity, or, in other words, how far would gravitation cause a body moving freely to fall in one second?

In this case, the whole weight moved by the force of gravitation on the bar, is $31\frac{1}{2}+31\frac{1}{2}+1=64$ oz. It is clear that 64 oz. will move only $\frac{1}{64}$ as far in one second as 1 oz. moved by the same force freely. Hence, $3\times64=192$ inches, would be the distance the bar would fall freely in one second. This distance is equal to 16 ft. When the experiment is accurate, at the level of the sea, it is found to be $16\frac{1}{12}$ ft.

§ 2. MOTION PRODUCED BY MORE THAN ONE FORCE.

(25.) If a body be acted upon by two forces which, separately, would cause it to describe the adjacent sides of a parallelogram, they will be equivalent to a single force, causing it to move through the diagonal of the parallelogram.

Hence the effect of two forces may be found by representing them by the two sides of a parallelogram, and then drawing the diagonal.

1. *If a body be acted on by two forces.*—Forces seldom act singly. It is by the combined action of at least two, often of more, that almost every motion is produced. The action of two forces may be illustrated by a very simple experiment. Place a ball at one corner of the table. Snap it with the fingers, lengthwise of the table, and it will roll along the side; or snap it across the table, and it will roll across the end. But skillfully snap it both ways at the same time, using both hands for the purpose, and it will roll in neither of these directions, but will move obliquely across the table.

The same thing is true of the action of natural forces, such as wind and tide. A ship, driven south by a direct wind, may at the same time be drifted east by a tide moving eastward. If so, it will, at every moment, be moving south and east, or in a straight line toward the southeast.

2. *Acting along the adjacent sides of a parallelo-*

gram.—The conditions of the motion of both the ball and the ship may be represented to the eye. Let A, Fig. 33, represent the original place of the ball or the ship. Suppose that while one force, if acting alone, would move the body to B, the other, if acting alone, would move it to D, *in the same time;* then when both act at once, the body will neither go to B nor D, but will go along the diagonal line A C, and will reach the point C in the same time it would have taken to go to either B or D.

Fig. 33.

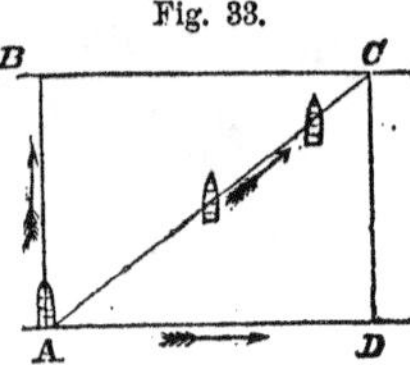

3. *They are equivalent to a single force.*—The two forces, acting in the directions A B and A D, produce a single motion along the line A C. A single force acting in the direction of A C would have produced the same effect. Hence two forces, acting in directions of the sides of the parallelogram, are *equivalent* to a single force acting in the direction of the diagonal.

The separate forces are called *components;* the single force which would produce the same effect, is called the *resultant*, and the process of finding the resultant is called the *composition of forces.*

4. *The resultant of forces may be found.*—The resultant of two forces may be found by representing them by two adjacent sides of a parallelogram and then drawing the diagonal. The lengths of the lines represent the strength, or the *intensity* of the forces.

In the case of the ship, for instance: suppose the wind able to drive it 10 miles, while the tide can drift it 5 miles. To find the actual path of the ship, draw the line A B, Fig. 33, to represent the 10 miles, and then the line A D, at right angles to it, and *one half as long*, to represent the 5 miles. Draw the lines B C and C D, to complete the parallelogram, and then draw the diagonal A C. This line represents the resultant of the two forces.

If more than two forces act at once, the resultant of all may be found by repeating the process. Find the resultant of two of them first; then compare this resultant and a third force; this second resultant and a fourth force; and so continue until all the forces have been used; the last resultant will be the resultant of all the forces.

(26.) Any force may be resolved into two others, which, acting together, would produce the same effect. This is done when we wish to know what part of a given force can be made available in a direction different from that in which it is exerted.

1. *A force may be resolved.*—To find the components of a given force, we may represent it by a line, and make this line the diagonal of a parallelogram; the adjacent sides of this parallelogram will represent the components. More than one parallelogram can be drawn on the same diagonal; so more than one set of components may be found for a single force.

The process of finding the components of a single force, is called the *resolution of forces.*

2. *To find the component which acts in a given direction.*—When a ball is thrown obliquely against the floor, it acts upon it with less force than when thrown perpendicularly against it. But a part of the force will still be exerted perpendicularly to the floor.

To illustrate this important point, let a ball A, (Fig.

34), be thrown against the floor, striking it at C. We may let the line A C represent the force with which the ball is thrown. Now construct the parallelogram, by drawing the lines A B and C D perpendicular to the floor, and then A D parallel to it. The lines A B and A D represent the components of the force A C.

Fig. 34.

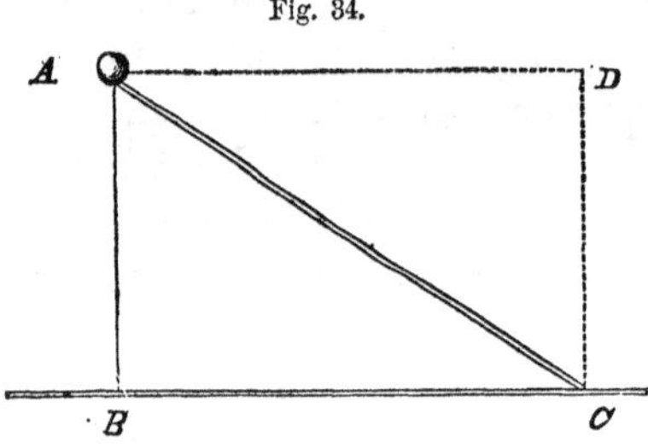

The line A B represents the amount of force exerted *perpendicularly* to the floor. To make the illustration more specific, we will suppose that, measuring the lines A C and A B, we find the latter to be $\frac{3}{5}$ as long as the former; if so, then the force exerted perpendicularly to the floor, will be $\frac{3}{5}$ of the force with which the ball is thrown.

To find the component which acts in *any* given direction, we may represent the original force by a straight line, and make it the diagonal of a parallelogram, one of whose adjacent sides is in the direction given. This side will represent the force required.

(27.) Two forces may act upon different points of a body in the same direction: their resultant will be equal to their sum. The point of the body to which this resultant is applied, will be as many times nearer to the greater force than the smaller one, as the greater exceeds the smaller in intensity.

The weight of a body is only the resultant of a set of parallel forces acting upon it in the same direction: and what is called the center of gravity, is its point of application.

1. *Two forces in the same direction.*—When two forces act upon a body in the same direction, they produce the same effect as a single force equal to their sum. If two horses, for example, draw a carriage, one with a force of 200 lbs., and the other with a force of 300 lbs., it is clear that a single horse, exerting a force of 500 lbs., would produce the same effect.

2. *The point of application.*—But if a single force is to take the place of two others, and produce exactly the same motion as they would when acting together, at what point of the body shall it be applied?

Suppose the body represented by A B (Fig. 35), to be acted upon by two forces, represented by the lines *c* and *d*, one just half the length of the other, the lesser force being 25 lbs., the greater, 50 lbs. Then the line *r*, just as long as both together, will represent the resultant, a force of 75 lbs.

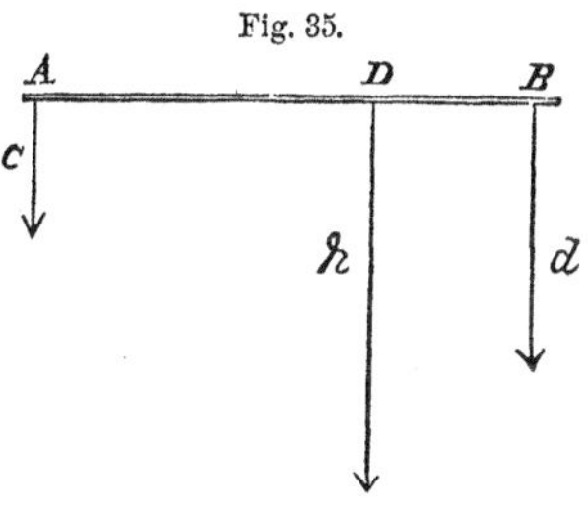

Fig. 35.

Now, if this resultant is to move A B, exactly as the two components would, it must be applied at some point, D, as many times farther from A than from B, as the force at A is times less than that at B. Since *c* is just half of *d*, the distance A D must be just twice as great as B D.

3. *The weight of a body.*—A body falling freely, is an example of motion caused by the action of parallel components. For, since the force of gravitation acts

upon *every molecule* of the body, we may regard the entire force as made up of as many separate forces as there are molecules. The sum of all these components is their resultant, and the value of this resultant is the *weight* of the body.

4. *The center of gravity.*—The point of application of this resultant, is the center of gravity. The center of gravity is usually defined to be—*that point in a body which, if supported, the body will rest in any position.* One can balance a book on the tip of his finger: the tip of the finger must be exactly under the center of gravity of the book. This point being supported, the whole body will rest.

The center of gravity is the exact middle point of a body of uniform density; it is toward the heavier side of one that is not. (See Silliman's Physics, pp. 39 to 45.)

A vertical line drawn through the center of gravity is called the *line of direction*, because it shows the direction a body will take when allowed to fall. That a body may stand upon a plane surface without falling, the line of direction must pass through its base. One body stands more firmly than another, only because it is more difficult to throw its line of direction beyond its base. A load of hay is easily overturned, because, the center of gravity being high, the line of direction may be easily thrown outside the base. A load of stone, having no greater weight, stands firm, because, the center of gravity being low, the line of direction can with difficulty be thrown beyond its base.

Animals instinctively incline their bodies always in such way as to keep their center of gravity over the space between their feet. Especially is this true of man.

A body is tottering in proportion as it has great height and a narrow base; but it is the prerogative of man to be able to support his commanding figure, erect and firm, under constant changes of position, over the very narrow base occupied by his feet.

(28.) Curved motion is produced by the action of at least two forces, one of which is a constant force, the other may not be.

The motion of a projectile is caused by the constant force of gravitation, and the impulse by which it is thrown.

1. *Curved motion.*—Whoever watches the varied and beautiful motions in nature, will find that they all take place in curves. In the ripples of the lake and the billows of the sea, he will see a wonderful variety of curved motions. The winds, and the clouds they carry, move in curves. Every swaying branch and leaf, and every nodding stalk of grass, moves in a curve.

2. *Is produced by at least two forces.*—The motion of a ball when fastened to the end of a string and whirled around the hand, is an example of curved motion. It is produced by the action of two forces. The impulse of the hand H (Fig. 36), which starts the ball, would, if it could act alone, carry it in a straight line from A toward B. But the string H A, held firmly by the hand, is a constant force which pulls it away from that path. The resultant of these two forces is represented by the circumference A B C D.

Fig. 36.

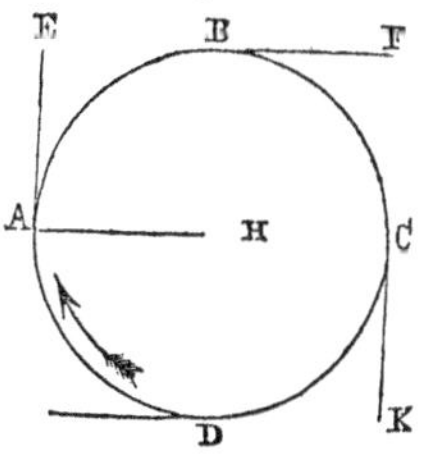

3. *One of which is constant.*—In the example just given, the force of the hand is an *impulsive* force; that of the string a *constant* force, and a curved motion is the result. Two impulsive forces will cause motion in a straight line; two *equal* constant forces will do the same. Two constant forces that are unequal will cause a curved motion; one, at least, of the forces must be constant.

The two forces which cause curved motion are called *central forces.* One of them alone, acting upon the ball at B (Fig. 36), would carry it along the line B F, or if the ball has reached C, would move it toward K. The influence of this component is to move the ball in a line which is tangent to the circle in which it revolves. This force is called the *centrifugal force.* The other component, which prevents the ball from moving in a straight line from the center of motion, is called the *centripetal force.* A simple and pleasant experiment may be performed to illustrate the effect of centrifugal force. To the handle of a small pail, filled with water, tie a cord firmly. Grasp the cord and swing the pail, fearlessly, in a vertical circle over the head; the centrifugal force will overcome the force of gravity, so that not a drop of water will fall, even when the pail is bottom side up over the head.

Circus riders incline their bodies toward the center of the ring around which they ride, that the centrifugal force may not throw them from their horses. Carriages, in rapid motion around the corner of a street, are sometimes overturned by this force. But the most wonderful examples of the action of central forces, are seen in the majestic movements of the heavenly bodies. Their orbits are ellipses. The impulse which drives the planets forward is the centrifugal force, while the centripetal force is the attraction of the sun, which holds them in their orbits.

4. *Projectiles.*—Any body thrown into the air is a projectile. The stone from the hand, the ball from the gun, and the arrow from the bow, are familiar examples of projectiles.

5. *Their motion is due to two forces.*—Leaving resistance of the air out of account, the motion of a projectile is due to the action of,

1st. The *impulse*, which starts it on its journey; and, 2d, the constant force of gravity.

Let us suppose a cannon ball, shot from the point *a* (Fig. 37), to go in the direction *a b*. At the end of the first second, the impulse given by the gunpowder would have thrown the ball to some point, as that marked I. But gravitation will, at the same time, be pulling the ball toward the ground. Represent the effect of this force in the first second by the line *a* I′. The resultant of these two forces will carry the ball to the point *e*. During the next second, the impulse acting upon the ball at *e*, would carry it to *s*, just as far as it would in the first, but gravitation would, in the same time, move it downward to *d*; the resultant of these two forces will carry the ball to *f*. In the third second, the impulse would throw the ball from *f*

Fig. 37.

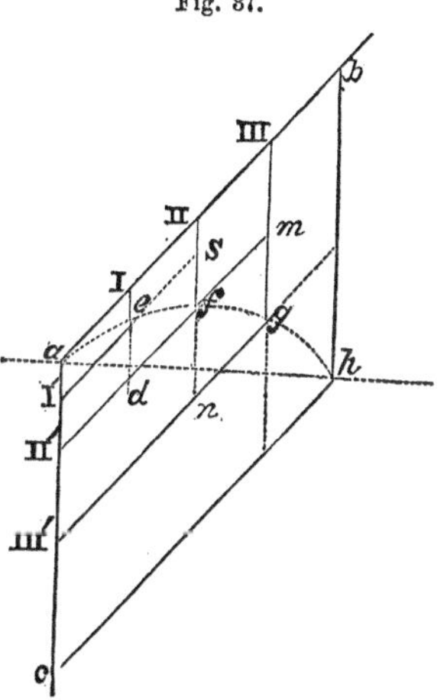

to *m;* but gravitation would pull it down to *n*, their joint action would carry it to *g*.

Now, let us remember that gravitation acts, not only at the beginning of each second, as the figure represents it, but also at *every other instant*, so that the path of the ball will bend, not at the points *e*, *f*, and *g* alone, but at every point between these, thus forming a curve reaching the ground at *h*.

The horizontal distance, *a h*, is called the *range* or the *random* of the projectile.

This distance depends upon the force applied to the projectile, and the angle at which it is thrown. Theory requires that the random be greatest when the projectile is thrown at an angle of 45° ; but the resistance of the air very much modifies the motion, so that, in practice, the greatest range is obtained at an angle much below 45°. The greatest range of an arrow is when the angle is about 36°.

The science of gunnery rests upon the laws of projectiles. The most skillful gunner is he who can most accurately, under all circumstances, compare and combine the forces of gunpowder, gravitation, and the resistance of the air.

§ 3. THE INDESTRUCTIBILITY OF FORCE.

(29.) Force, like matter, is indestructible. Whatever force has acted to put a body in motion, the same amount must be exerted by the moving body upon others, before it can come to rest.

Three well-marked cases are before us :—

1st. In which a body moves without resistance from other bodies ;

2d. In which a body, moved by an impulsive force, meets with resistance;

3d. In which a constant force is applied to overcome the resistance.

1. *Force is indestructible.*—It was once thought that bodies of matter could be destroyed. It seems so yet to a careless observer; but when he has learned how to search for their scattered fragments, he finds that every atom still exists. Forces likewise vanish; but when the motions they produce have been changed to rest, and after every trace of their action seems to have been lost, they have been chased from their hiding-places, until it is proved that every impulse still acts—that while it may change from form to form, and show itself in a multitude of ways, yet not a single impulse of force can be destroyed.

2. *Motion can not cease without exerting the same amount of force which produced it.*—The force of gunpowder is expended in giving motion to a ball: the ball exerts the same force upon whatever obstacles it meets. A small force will give slow motion; the body moving slowly, will, on meeting an obstacle, exert the *same* small force. A greater force will give a swifter motion; the body moving swiftly will strike another with the same greater force. Thus a bullet may simply bruise an arm, or it may pierce a tree, or shatter a block of the hardest stone, according to the velocity with which it strikes; and the velocity will, in turn, depend upon the force which puts the ball in motion.

3. *Suppose a body move without resistance.*—If a body should move without any resistance to its

motion, until it suddenly strikes an obstacle, the force with which it would strike, would be exactly equal to that which gave it motion. If a force of ten pounds puts a body in motion, it would hit the other with a ten-pound force. The force which a body moving without resistance can exert upon an obstacle is called its *momentum.*

Suppose a body, weighing 1 lb., move with a velocity of 1 ft. a second. If it meet with no resistance, it will strike another body with a certain force, or momentum, which we will call 1. The momentum of a 2 lb. weight, with the same velocity, would be twice as great; it would be 2. A weight of 1 lb. moving twice as fast, would also have twice the momentum of the first; it would be 2. A weight of 3 lbs., would have a momentum of 3, and then, if its velocity be doubled, it would, on that account, have twice as much momentum; it would be 6. The momentum of a moving body is thus seen to vary with its *weight* and its *velocity*, and we find that *the momentum of a body is equal to the product of its weight multiplied by its velocity.*

4. *Suppose motion due to an impulse meet with resistances.*—When a moving body meets with the resistance of air, of friction, or of any other influence, before it strikes another, the force which started it will be exerted, partly upon this resistance, and partly upon the object which it finally strikes; but *the sum of these two parts* will exactly equal the force which set the body moving. Thus a stone thrown from the hand will strike a tree at a distance with much less force than it receives, but in its motion it has been compelled to move the air before it, and if the force which it exerts in this way, be added to that which it exerts on the

tree, the aggregate will be just equal to that which the hand exerted upon the stone at first.

So, if a ball be rolled upon the ground, its force must act upon the air in front of it, and upon the roughness of the ground under it: all its force is thus used, and it stops; but the total amount gradually expended in this way, must just equal the sudden impulse which started the ball upon its journey.

5. *Suppose a constant force applied to overcome resistances.*—When a constant force is applied to keep the body moving uniformly, in spite of resistances, the force with which it strikes an obstacle must be equal to that which acts to keep it moving. Thus, if a train of cars, kept in motion at the rate of 20 miles an hour, suddenly strike another train going with equal velocity in the opposite direction, the entire force of steam, expended to keep up the motion of both trains, will be suddenly exerted to dash them both to pieces.

This force, which a body, moving against resistances with a uniform motion kept up by a constant force, will exert upon an obstacle, is called *living force*, or *vis viva.*

Now, if the velocity of a body be doubled, the force required to keep it uniform will be four times as great. Take the case of a ship moving through water. If the velocity of the ship be doubled, twice as much water will have to be moved in the same time; it will take a double force to do this. Moreover, every particle of water will have to be moved twice as fast; it will take a double force to do this also. To do both these things at the same time, the force must be twice doubled, or made four times as great. This is true of all resistances, and for all velocities. In other words, the force required to keep a body moving against resistances, will be in proportion to the *square of the velocity.*

But when, under such circumstances, the moving body strikes an obstacle, the force which keeps it moving, must be suddenly exerted upon the body struck. It thus appears that the *living force* of a body is in proportion to the square of its velocity. It *is equal to the product of the weight multiplied by the square of the velocity.*

§ 4. OF MACHINERY.

(30.) The principle of momentum applied to any one of the simple machines, will determine its law of equilibrium.

1. *The principle of momentum.*—Momentum has been defined to be the *force* which a moving body, meeting no resistances, can exert. Now, two bodies exerting *equal forces* upon each other in opposite directions, will, when at rest, just balance each other, or be in *equilibrium.* This principle is called the *principle of momentum.* It states briefly that *two forces, in opposite directions, will be in equilibrium when their momenta are equal.*

Let us illustrate a single case by means of Fig. 38.

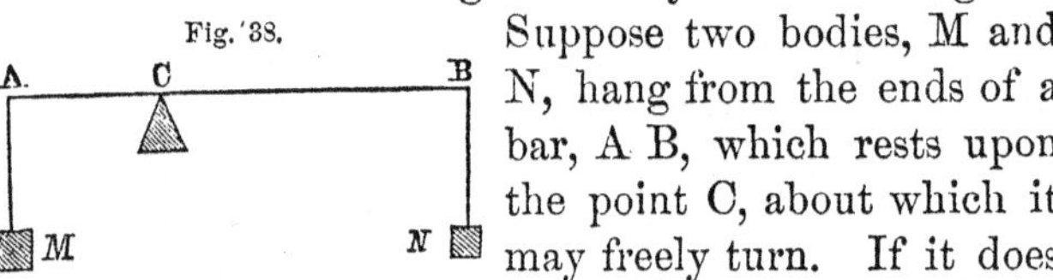

Fig. 38.

Suppose two bodies, M and N, hang from the ends of a bar, A B, which rests upon the point C, about which it may freely turn. If it does turn, and M goes up, N will go down, and if the distance B C is twice the distance A C, then N will go twice as fast as M. In all cases, the velocities of the two bodies will have the same ratio as their distances, A B and B C, from the center of motion. *These lines may then be taken to represent velocities.* Then the momentum of M will be M × A C, and that of N will be N × B C. Now if these momenta are equal, then the two bodies will be exerting equal forces upon the bar A B, and if once brought to rest, they will just balance each other.

2. *Machines.*—Machines are instruments by which forces may be applied to overcome resistance, or do work. They are so made that a small force, by moving rapidly, may overcome a greater resistance, or a great force, by moving slowly, may put a small resistance in rapid motion. In all cases the momenta of the two forces must be equal.

The resistance to be overcome is always called the *weight:* the force which overcomes it is called the *power.*

3. *Simple machines.*—There are six simple forms of machines, usually called the mechanical powers. Out of these six simple machines all forms of machinery, complex as they may be, are made. We name them in the order which is to be followed in describing them.

1. The Lever.
2. The Wheel and Axle.
3. The Pulley.
4. The Inclined Plane.
5. The Wedge.
6. The Screw.

4. *The law of equilibrium.*—By the term, law of equilibrium, is meant a statement of the relation which must exist between the power and the weight, in order that, when at rest, they may just balance each other.

(31.) Levers are of three classes. The principle of momentum applied to the lever, shows that the power and weight will be in equilibrium when they are to each other inversely as the perpendicular distances from the fulcrum to the directions in which they act. A compound lever acts on the same principle. Applications of the lever are very numerous.

Fig. 39.

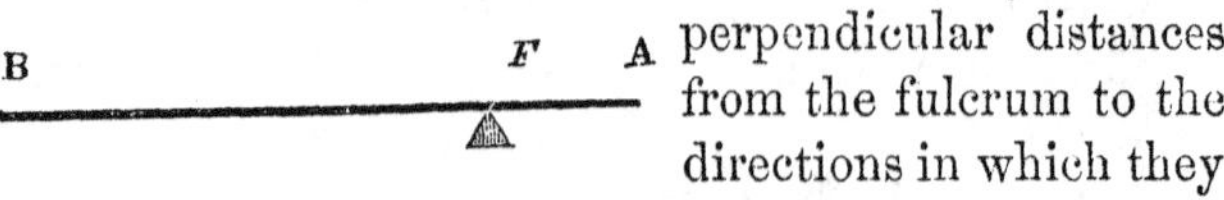

1. *Levers.*—A lever is an inflexible bar, able to turn freely upon one point. Thus, if the line A B (Fig. 39) represents an inflexible bar, resting upon some support at F, upon which it has free motion, it represents a lever. The point F, about which the lever turns, is called the *fulcrum*.

Fig. 40.

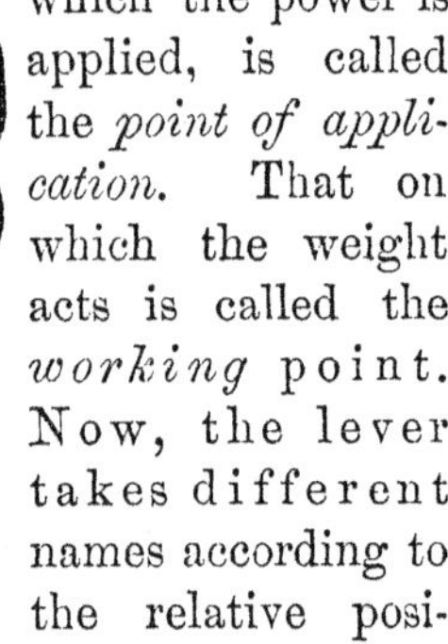

2. *Three classes of levers.*—That point of a lever to which the power is applied, is called the *point of application*. That on which the weight acts is called the *working* point. Now, the lever takes different names according to the relative positions of the point of application, the working point, and the fulcrum. In the lever represented in Fig. 40, whose fulcrum is at F a power (P) acts upon one

Fig. 41.

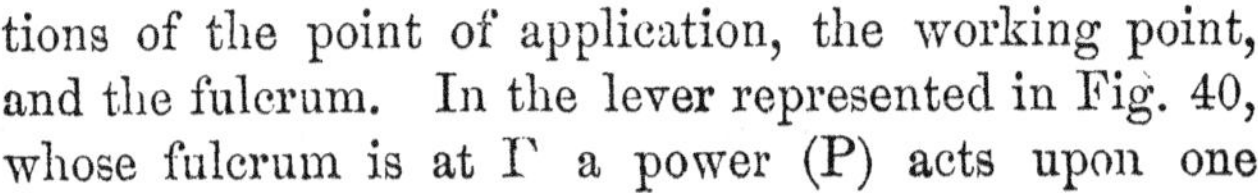

end of the lever (A) while a weight (W) acts upon the other (B). The fulcrum is between the point of

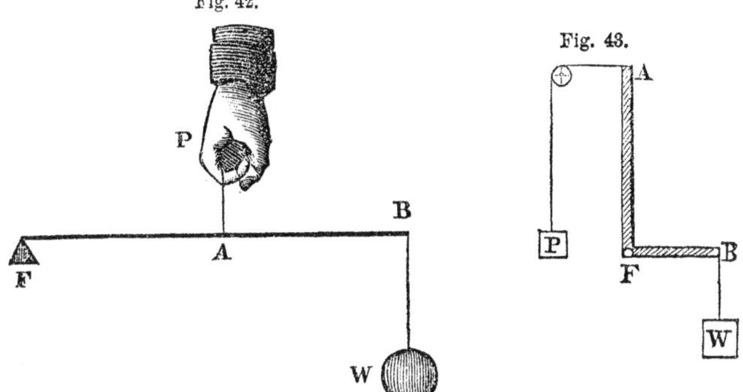

Fig. 42. Fig. 43.

application and the working point. This is called a lever of the *first class.*

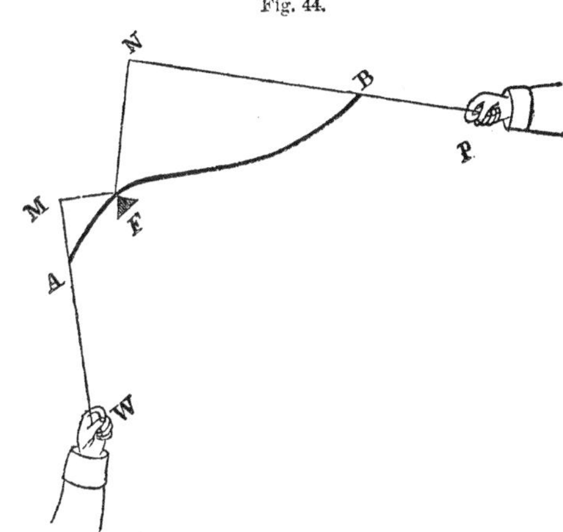

Fig. 44.

In the lever, Fig. 41, the working point B, is between the point of application A, and the fulcrum E. This is a lever of the *second class.*

In the lever, Fig. 42, the point of application (A) is between the working point (B) and the fulcrum, F. This is a lever of the *third class.*

All levers belong to these three classes. They need not, however, be made in the simple straight form shown by the figures. In Fig. 43, the line A F B represents a lever whose arms make a right angle at the fulcrum, F. It is a lever of the first class; so also is that shown in Fig. 44, by the curved line A F B.

3. *Application of the principle of momenta.*—Now if we examine the figures which represent the three classes of lever, we see that in each one, the power (P) and the weight (W) are two forces which act in opposite directions. They will be able to just balance each other, when of such strength that, when moving, their *momenta are equal.* The lines B F and A F represent their velocities [see (30.) 1]. The momentum of the power is therefore $P \times AF$; that of the weight is $W \times BF$. If equilibrium takes place only when the momenta are equal, then

$$P \times AF = W \times BF; \text{ hence,}$$
$$P : W :: BF : AF.$$

This proportion teaches us that the power and weight will be in equilibrium, when they are to each other inversely as the distance of their points of application from the fulcrum.

It may be, however, that the power and weight do not act perpendicularly upon the lever. This case is represented by Fig. 44. The lever A B has its fulcrum at F. The power (P) and the weight (W) act obliquely at B and A. Now it is evident that the force of the power, acting obliquely at B, is not all expended to lower the lever [see (26.) 2], but that if it were acting upon the point N, perpendicularly, it would exert all its force to move the arm N F. So the effect of the weight acting obliquely upon A, will be the same as if it were acting perpendicularly upon an arm, M F. Hence, $P \times N F$ may be taken as the momentum of the power, and $W \times M F$ as the momentum of the weight. Putting these momenta equal,

$$P \times N F = W \times M F; \text{ hence,}$$

$$P : W :: M F : N F.$$

This proportion teaches that *the power and weight will be in equilibrium, when the power and weight are inversely as the perpendicular distances from the fulcrum to the directions in which they act.*

This principle is called the law of equilibrium for the lever. It will apply to all possible forms.

4. *The compound lever.*—In a compound lever several simple levers are generally so fixed, that the short arm of one may act upon the long arm of another. Fig. 45 shows a compound lever made up of two simple levers having their fulcrums at F, and F′.

Fig. 45.

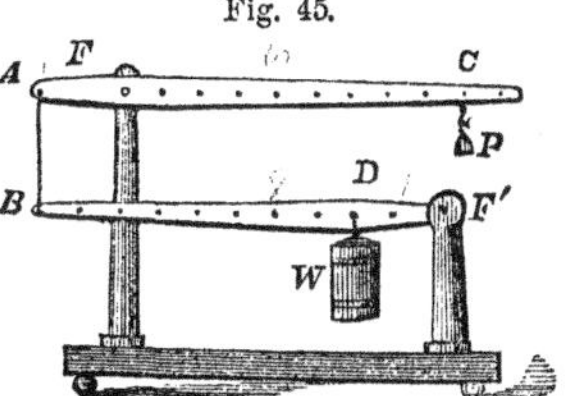

In this case the momentum of the power will be equal to $P \times C F \times B F'$, and that of the weight will be equal to $W \times D F' \times A F$. If these products are put into the form of an equation it will be seen that the power and weight will be in equilibrium when the power multiplied by the product of all the arms on its side of the fulcrum, is equal to the weight multiplied by the product of all the arms on its side.

The compound lever is used when a small force is required to sustain a large weight, and it is not convenient to have a very long lever. If the long arms of the simple lever be 6 and 8 ft., and each short arm is 1 ft., then 1 lb. power at C, will balance 48 lbs. at D; while if a simple lever had been used whose long arm was as long as those two long ones together, 6 + 8 = 14 ft., and whose short arm was 1 ft., then 1 lb. at C, would only be enough to balance 14 lbs. at D.

5. *Applications of the lever.*—Of levers of the first kind many familiar examples might be named. The handspike and crow-bar are levers of this class. Shears and pincers are pairs of levers, also of the first class; their fulcrums being at their joints.

The *balance* is one of the most useful applications of the lever. Fig. 46 represents this instrument.

Fig. 46.

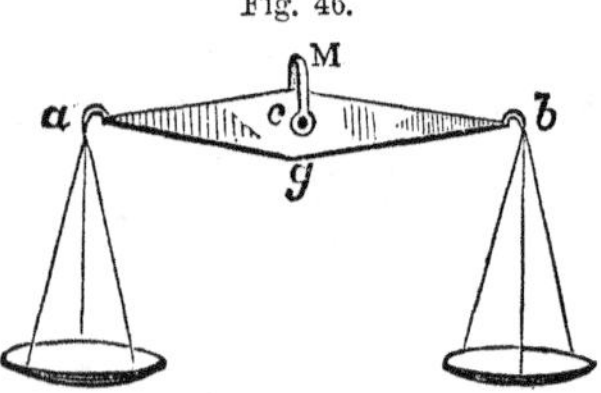

The beam *a b* is a lever poised at its centre, the pivot or fulcrum *c* being a *little above* its center of gravity. From the ends of the beam the scale pans are hung, in one of which is put the body to be weighed, and in the other, the weights of metal to balance it. Balances are of continual use in commerce; they are indispensable in the laboratory

Fig. 47.

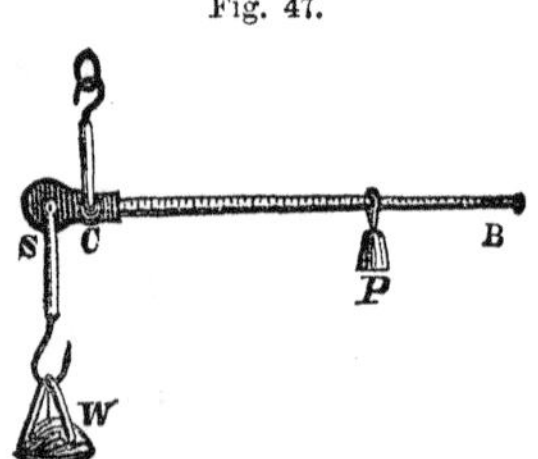

of the chemist, for whose use they are made with so great skill that a weight equal to the $\frac{1}{10,000}$ of a grain can be easily weighed.

The *steelyard* is also a lever of the first class, but with unequal arms. The body W, Fig. 47, to be weighed, is hung from the short arm of the lever S B, and it is balanced by a small weight, P. It is clear that this small weight will balance more weight in the body W, as it is moved farther and farther from the fulcrum C. The arm B C has notches cut upon it, and numbered, to denote the pounds or ounces in W, balanced by P, when at these points.

Levers of the second class are not so common; the *wheelbarrow*, however, is an example sufficiently familiar. The axle of the wheel is the fulcrum, to the opposite ends of the handles the power is applied, while the load, or the weight, rests between these points. The oar of a boat is a lever of this kind, where, singularly enough, the unstable water serves as a fulcrum; the hand is the power at the other end of the lever, while the boat is the weight between them.

Levers of the third class are often met with in the arts. The common fire-tongs and the sheep-shears are pairs of levers of this kind. Their fulcrums are at one end; the resistance to be overcome is put between their parts near the other end, while the fingers, which afford the power, are between the fulcrum and the weight.

(32.) The wheel and axle acts on the principle of a lever. The power and weight will be in equilibrium when the power is to the weight as the radius of the axle is to the radius of the wheel.

A compound wheel and axle acts on the same principle as a compound lever.

One wheel may be made to turn another by friction, by cogs, or by bands.

Applications of this machine are common and important.

1. *The wheel and axle.*—One form of the wheel and axle is shown in Fig. 48. It consists of a wheel (B) firmly fastened to an axle (A), and turning freely around an axis, one end of which is shown at C. The power (P) acts upon the circumference of the wheel, and the weight (W) acts upon the axle by means of a rope winding around it in the opposite direction.

2. *It acts on the principle of the lever.*—If we have an end view of the machine, it will be seen, as shown in Fig. 49, where the large circle represents the wheel, and the small circle, the axle; the center C, being the end of the axis. At the point A, the power

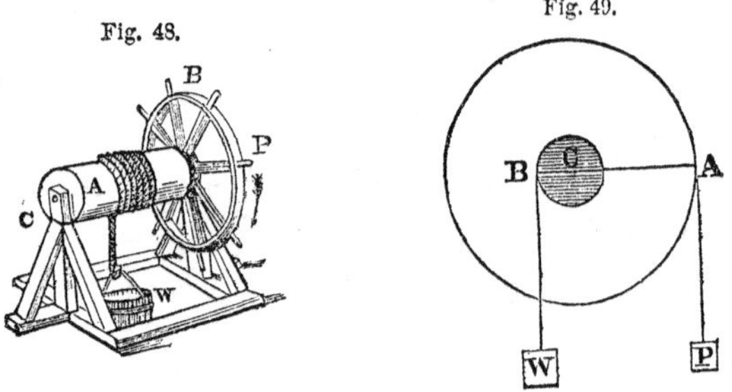

Fig. 48. Fig. 49.

acts on the wheel, and from the point B, on the other side of the center, the weight is suspended. Now, if a straight line A B join the points A and B, it will pass through the center, and represent a lever, whose fulcrum is at C. It is upon the ends of such a lever that the power and weight are constantly acting.

3 *Application of the principles of momentum.*—The momentum of the power is $P \times A\,C$; that of the weight is $W \times B\,C$. If the two forces are able to balance each other, these momenta are equal. Hence,

$$P \times A\,C = W \times B\,C: \text{ or,}$$
$$P : W :: BC : AC.$$

But, in the figure, we notice that A C is the radius of the wheel, and that B C is the radius of the axle, Then the proportion teaches us that *the power and weight will be in equilibrium when the power is to the weight, as the radius of the axle is to the radius of the wheel.*

If the radius of the axle is 1 ft., and that of the wheel 3 ft., then 1 lb. will balance 3 lbs.

4. *The compound wheel and axle.*—When more than one wheel and axle are connected, so that the axle of each may act on the wheel of the next, the machine is a compound wheel and axle. Such an arrangement is shown in Fig. 50. We may get the law of equilibrium in the same way as in the compound lever. The momentum of the power will be P, multiplied by the several radii of the wheels; that of the weight will be W, multiplied by the several radii of the axles. If the two forces are able to balance each other, these values must be equal. Hence we learn, that in a compound wheel and axle, the power and weight will be in equilibrium, when the power multiplied by the product of the radii of the wheels, equals the weight multiplied by the product of the radii of the axles.

Fig. 50.

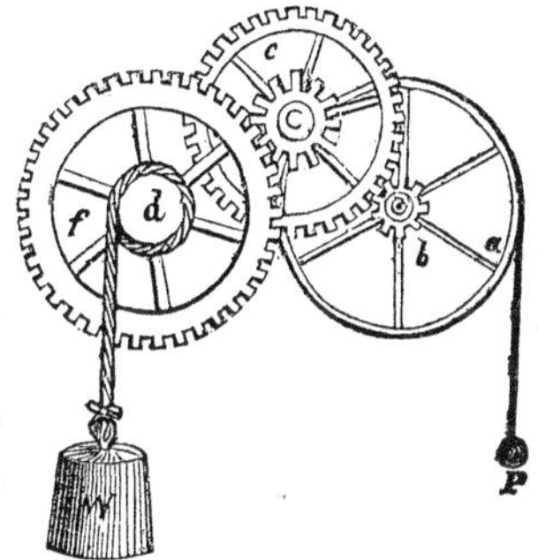

It is easy to see that, in this way, a small power may be made to balance a much larger weight than it could by acting upon a simple wheel and axle, unless the wheel should be so large as to be unwieldy.

5. *One wheel may turn another by means of cogs.*—In Fig. 50, there may be seen projecting teeth on the circumferences of the axles, *b* and *c*, which fit into equal notches on the circumferences of the wheels. Neither the axles nor the wheels can turn without causing the other to turn also. This is the common and convenient method of giving motion from one wheel to another. The wheels of a clock are cog-wheels: those of a watch also beautifully illustrate this mode of communicating motion.

6. *By friction.*—When the circumferences of the wheels and the axles are made smooth, they may be pressed so snugly together, that neither can turn without turning the other at the same time, in the opposite direction. In this case, the motion is communicated by the *friction* of the parts against each other.

7. *By bands*—A third method of giving motion to a train of wheel-work, consists in the use of bands or belts, which encircle the parts which are to act upon each other. In the spinning-wheel, for example, the spindle is turned by a band which passes around it and the axle of the wheel-head. Another band passes around the wheel-head and the large wheel, which is turned by the hand of the spinner. From the horse-power of a thrashing-machine, also, motion is given to the cylinder by means of a belt.

8. *Applications of the wheel and axle.*—Many forms of the wheel and axle are in common use: the windlass is one of the most familiar, being often used to raise water from wells. One form of the windlass is represented in Fig. 48. A crank is often used in place of the wheel, B. The common grindstone is a homely illustration of the wheel and axle: the crank is in place of

a wheel; the stone itself is the axle. The power is the force of the hand, while the weight is the resistance offered by the tool pressing on the edge of the stone.

If the axle is in a vertical position, and the forces of power and weight act horizontally, the machine is then called a *capstan*, and is much used on board of ships.

The compound wheel and axle is used in almost every mill and factory. Two objects are sought in its use: either great resistance is to be overcome, or rapid motion is to be secured. To overcome great resistances, the power is applied to the circumference of the first wheel in the system, and the weight is acted upon by the last axle. This case is shown in Fig. 50. To secure rapid motion, the power is applied to the first axle, while the weight is acted upon by the circumference of the last wheel. The same figure illustrates this case also, if we will suppose the heavy body W, to act as a power to put the lighter body P, in motion. If we suppose the radius of each axle to be 1 ft., and of each wheel 10 ft., then $P \times 10 \times 10 \times 10 = W \times 1 \times 1 \times 1$: or, $1{,}000\,P = W$. Now, W being 1,000 times heavier than P, P must move 1,000 times faster than W. In this way, a great power may be changed into rapid motion. An example of this is found in the saw-mill where the slow motion of a heavy body of water, acting against a water-wheel, is given, by means of cogs and belts, from wheel to wheel, until it reappears, multiplied a thousandfold, in the buzzing saw.

(33.) The pulley may be either fixed or movable. In the fixed pulley the power and weight will be in equilibrium when they are equal.

In the movable pulley, with a single rope, the

power and weight will be in equilibrium when the power is equal to the weight, divided by the number of branches of rope which sustains the weight.

In movable pulleys, with separate ropes, the power and weight will be in equilibrium when the power equals the weight, divided by 2, raised to a power, shown by the number of pulleys.

The applications of the pulley are common and important.

1. *The pulley.*—A pulley is a grooved wheel, turning freely about its axis, with a rope passing over or around it. It is shown in Fig. 51. The grooved

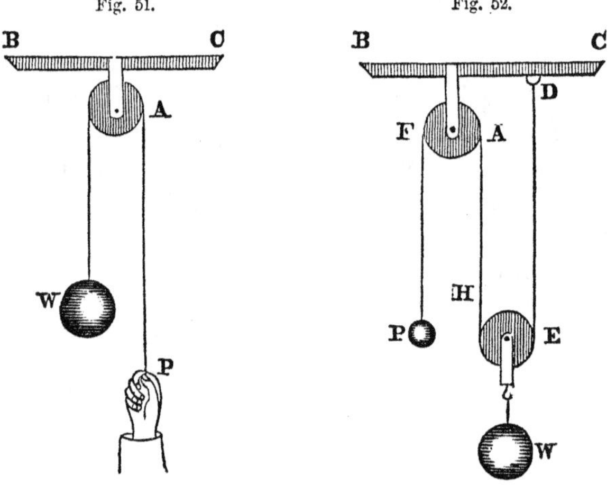

wheel A, moves freely upon its axis, while over its circumference goes the rope, to the ends of which the power and the weight are fastened.

2. *Is either fixed or movable.*—If the axis of the pulley is stationary (see Fig. 51), the pulley is called a

fixed pulley. The *movable pulley* is one whose axis moves with the weight. This will be understood by means of Fig. 52. The wheel E, is a movable pulley. From its axis the weight is hung, while the rope, one end of which is fastened to a fixed support at D, passes under it and then over a fixed pulley A. The power is applied to this end of the rope.

3. *The principle of momentum applied to the fixed pulley.*—The fixed pulley is shown in Fig. 51, to which we again refer. It is clear that, when motion occurs, the power (P) will go down with exactly the same velocity as the weight (W) goes up. To have equal momenta when the velocities are equal, the bodies must have equal weights. Hence, *in the fixed pulley the power and weight can balance each other only when they are equal.*

Fig. 53.

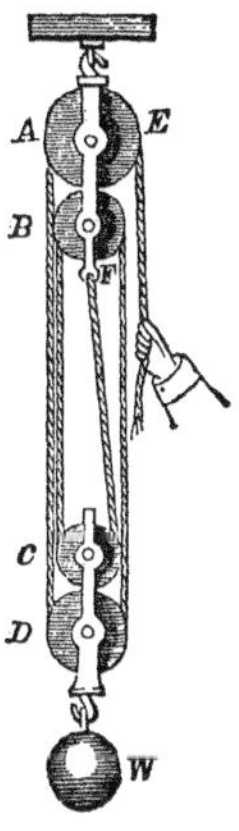

4. *The principle of momentum applied to the movable pulley with a single rope.*—In the movable pulley, with a single rope (see Fig. 52), the weight rests upon two branches of the rope, H and E, and when it rises, both branches must be equally shortened. But the rope F P will lengthen just as much as both the branches shorten. The power (P) moves downward just twice as fast as the weight (W) goes up. Let V represent the velocity of the power, then $\frac{V}{2}$ will represent the velocity of the weight, The momentum of P will be $P \times V$, and that of W will be $W \times \frac{V}{2}$, and the two forces can balance when

$$P \times V = W \times \frac{V}{2}, \text{ or, when } P = \frac{W}{2}.$$

Now let us take another case. Suppose there are

two movable pulleys, C and D (Fig. 53), with a single rope, one end being fastened at F, while to the other end the power P, is applied. In this case we find that the weight is supported by *four* branches of the rope, and we see, too, that when it rises, all four of these branches must be shortened alike. But the rope, E P, must at the same time lengthen as much as all the branches shorten, so that the velocity of P downward must be four times as great as that of W upward. Then, if V is the velocity of P, $\frac{V}{4}$ will be the velocity of W; and, if their momenta are equal,

$$P \times V = W \times \frac{V}{4}, \text{ or, } P = \frac{W}{4}.$$

In like manner, if three movable pulleys are used, we should find that, to be in equilibrium, $P = \frac{W}{6}$.

If, now, we notice that in each of the values of P just found, the denominator of the fraction is the number of branches of the rope which supports the weight, we have this general principle: in movable pulleys, with a single rope, *the power and weight will be in equilibrium when the power equals the weight divided by the number of branches which support it.*

Fig. 54.

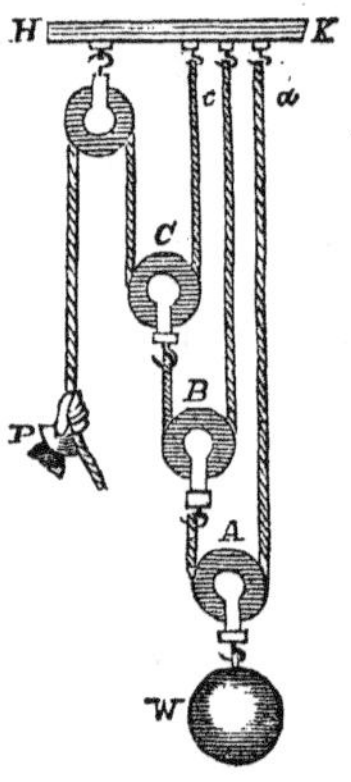

5. *The movable pulley with separate ropes.*—When each pulley has a separate rope, the law is very different. Fig. 54 shows such a system. The three ropes, *a b c*, are fastened to the beam H K. The first, after passing around the pulley A, is fastened to the axis of the one above: so the rope *b*, after going around the pulley B, is fastened to the axis of C: but the rope *c*, after going over the pulley C, passes over

a fixed pulley, and receives the power at the other end.

6. *The principle of momentum applied to the movable pulley with separate ropes.*—This system is only a *combination of movable pulleys with a single rope.* Suppose the pulleys A and B were taken away, the weight being hung from the axis of C, there would be left an arrangement exactly like that shown in Fig. 52. C is a movable pulley, with a single rope, to lift the pulley B. B is likewise a movable pulley, with a single rope, to lift the pulley A; while A is itself a movable pulley, with a single rope, to lift the weight W. No new application of momentum is needed. The effect of the power (P) will be doubled by each pulley thus:—

$$\text{With 1 pulley, } P = \frac{W}{2} = \frac{W}{2^1};$$

$$\text{" 2 pulleys, } P = \frac{W}{4} = \frac{W}{2^2};$$

$$\text{" 3 " } P = \frac{W}{8} = \frac{W}{2^3}.$$

If we notice that the denominator in each of these values of P, is a power of 2, whose *index* is the *number of pulleys*, we infer that, in a system of movable pulleys with separate ropes, the power and weight will be in equilibrium when the power equals the weight divided by a power of 2, whose index is the number of pulleys.

For example, with a system of 5 pulleys, how much weight will a power of 10 lbs. balance?

$$P = \frac{W}{2^5}; \text{ or } 10 = \tfrac{W}{32}; \text{ hence } W = 320 \text{ lbs.}$$

7. *Application of the pulley.*—No mechanical advantage is gained by the use of the *fixed pulley*, because the weight must move just as fast as the power, yet it

is of great value in the arts, for changing the direction of forces. A sailor standing upon the deck of his ship may, by the use of a fixed pulley, hoist the sail to the top of the loftiest mast; or when heavy bales or boxes are to be lifted to the upper floors of warehouses, a horse, trotting along the level yard or street (Fig. 55), will lift them as effectually as though he were able to climb the perpendicular wall with the same rapidity.

The *movable pulleys* with single rope, are in common use for moving heavy weights through considerable distances. Merchandise may be lifted by means of them, from the hold of a ship to the wharf, or to the upper stories of store-houses; or by a different arrangement of the machine, the ship itself may be drawn from the water for repairs. In practice, the fixed pulleys of a system are placed side by side, and thus form what is called a *block*: the movable pulleys, likewise side by side, form another block. By this means the system is made compact.

Fig. 55.

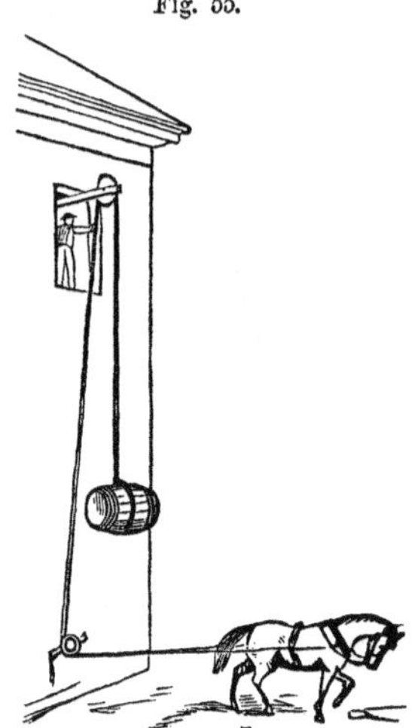

In all pulleys there is a loss of power, due to the friction of the pulleys in the blocks, to the weight of the lower block, and to the stiffness of the ropes used; so that the weight, actually overcome by a given power, is always less than the laws of equilibrium would afford.

(34.) The principle of momentum applied to the inclined plane shows :—

1st.—That, when the power acts parallel to the length of the plane, the power and weight will be in equilibrium when the power is to the weight as the height of the plane is to its length;

2d.—That, if the power acts parallel to the base of the plane, the power and weight will be in equilibrium when the power is to the weight as the height of the plane is to its base.

The applications of this machine are very numerous.

1. *The inclined plane.*—Any plane, hard surface, placed in an oblique position, may be used as an inclined plane. In Fig. 56, A B represents an inclined plane.

Fig. 56.

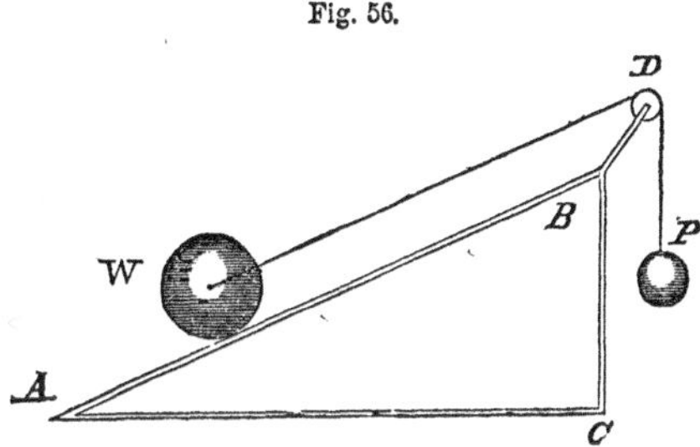

The distance B C, is the *height* of the plane, and A C is its base. The weight W, may be urged up the plane by a force acting parallel to A B, or parallel to A C, or at any angle to these. We are to notice the first two cases only.

2. *If the power acts parallel to the length of the plane.*—In the figure the power P, by means of a rope going over the fixed pulley D, at the top of the plane, acts upon the weight W, in a direction (W D) parallel to the length A B, of the plane.

Now, a force which will urge the weight from A to B,

is *lifting* it only through the *vertical* height, C B. But while the weight goes from A to B, the rope passing over the pulley, will let the power down a distance equal to A B, in the *same time.* The velocity of the weight may, therefore, be represented by the line C B, and the velocity of the power by the line A B. The momentum of the power is, therefore, P × A B; that of the weight, W × C B. When these momenta are equal, the two forces will be able to balance each other. Thus:—

$$P \times AB = W \times CB; \text{ or}$$
$$P : W :: CB : AB.$$

This proportion teaches us that, *when in equilibrium, the power is to the weight as the height of the plane is to its length.*

Fig. 57.

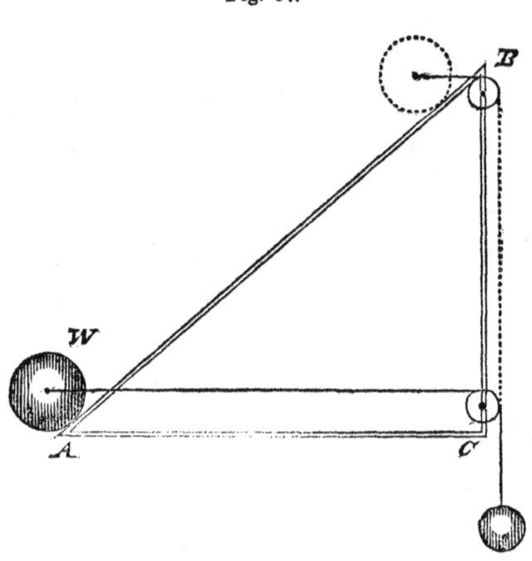

If, for example, the height C B, is 4 feet, and the length of the plane A B, is 16 feet, a power of 1 lb. will balance a weight of 4 lbs. For—

1 lb. : 4 lbs. :: 4 ft. : 16 ft.

3. *If the power acts parallel to the base of the plane.* —Let A B, Fig. 57, represent a plane whose height is C B, and whose base is A C. The power acts upon the weight by means of a cord passing over the pulley at C, in a direction parallel to A C. To move the weight from A to B, will be lifting it only through the *vertical* height, C B. If the pulley C, could be raised while the weight goes up, so as to keep the cord parallel to A C, then the cord, passing over the pulley, will let the power down a distance equal to A C. The line C B, represents the velocity of the weight, and A C the velocity of the power. The momentum of the power is therefore $P \times A\ C$, and that of the weight, $W \times C\ B$. If now, these momenta are equal, the power and weight will just balance each other. Hence—

$$P \times A\ C = W \times B\ C; \text{ or}$$
$$P : W :: A\ B : A\ C.$$

From this proportion we learn that, when the power acts parallel to the base of the plane, *the power and weight will be in equilibrium when the power is to the weight as the height of the plane is to its base.*

Thus, if the height of the plane is 2 ft. and the base is 10 ft., a power of 1 lb. will balance a weight of 5 lbs. For 1 lb. : 5 lbs. :: 2 ft. : 10 ft.

4. *Application of the inclined plane.*—This machine is used to lift heavy weights through short distances. Many familiar examples might be named. If a barrel of merchandise is to be placed upon a wagon, it is often rolled up on a plank, one end of which rests upon the ground, the other upon the wagon. A hogshead which a dozen men could not lift, may thus be loaded by the strength of one or two.

Our common stairs are, in principle, inclined planes,

the arrangement of steps only giving a firm footing. If the distance between the floors be $\frac{3}{4}$ the *length* of the stairs, then, besides the ordinary effort of walking, the person must continually, while going up, labor to lift $\frac{3}{4}$ of the weight of his body.

(35.) The wedge, in its most common form, is made up of two inclined planes joined together at their bases. The sharper the wedge, the greater the resistance which may be overcome by it.

1. *The Wedge.*—This instrument is shown in use by Fig. 58. A B is called the back of the wedge: A c and B c, are its sides, and c is its edge. It is generally used in cleaving timber, and sometimes for raising heavy weights through very short distances. For these purposes its edge is put into a crevice made for it, and it is then driven by blows with a sledge.

Fig. 58.

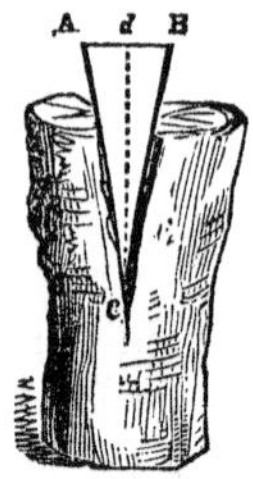

Since we can not calculate the force of a blow, no attempt will be here made to establish any law of equilibrium.

(36.) The principle of momentum applied to the screw, shows that:—

The power and weight will be in equilibrium, when the power is to the weight, as the distance between two contiguous threads is to the circumference of the circle in which the power moves.

The screw is used extensively to produce great pressure: it is often used to measure delicate distances.

1. *The screw.*—The screw consists of a cylinder of wood or metal, with a spiral groove winding around its circumference. This grooved cylinder (C, Fig. 59)

passes through a block N G, on the inside surface of which is a spiral groove, into which the raised parts of the cylinder exactly fit. The block is usually called the *nut.* The raised parts between the grooves of the cylinder are called the *threads.*

Suppose the nut to be stationary, then, if the screw is turned by a power acting upon the lever at B, it must advance downward at every revolution, and the pressure of the advancing screw will be exerted upon any object placed under the press-board, E F, against which the end of the screw presses.

Fig. 59.

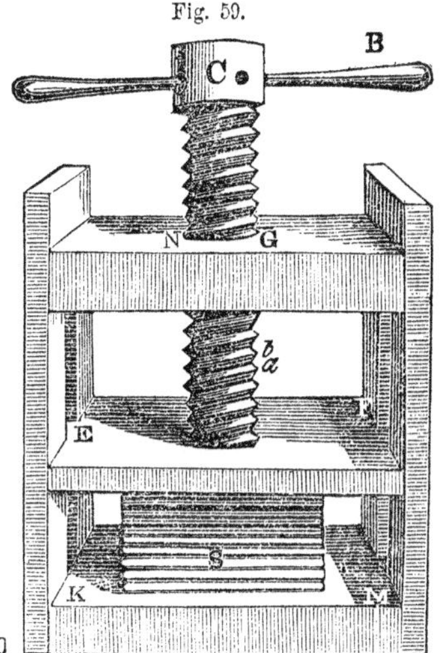

2. *Application of the principle of momentum.*—By one turn of the screw, it will advance downward a distance just equal to the distance between two contiguous threads. The press-board E F, which may be regarded as the weight, will be moved along through a distance equal to *a b*, by every turn. The power acting at B will, in the same time, move through the circumference of the circle whose radius is B C. These distances, through which the power and weight pass in the same time, may represent their velocities. Hence the momentum of the power will be P × Circumference of the circle whose radius is B C, and that of the weight will be W × *a b*. If these momenta are equal, the two

forces, when at rest, will be in equilibrium. Hence:—

$$P \times \text{Circ. } BC = W \times ab; \text{ or}$$

$$P : W :: ab : \text{Circ. } BC.$$

This proportion teaches that the power and weight will be in equilibrium when the power is to the weight, as the distance between two contiguous threads is to the circumference of the circle in which the power moves.

Thus, if the distance between the threads is $\frac{1}{2}$ in., and the circumference, traveled by the power, is 5ft., or 60 in., what weight on the nut, would 1 lb. power at B balance?

$$1 \text{ lb.} : W :: \tfrac{1}{2} \text{ in.} : 60 \text{ in. } W = 120 \text{ lbs.}$$

3. *Application of the screw.*—The screw is used when great weights are to be lifted short distances, or when heavy pressure is to be exerted. By its use, cotton is pressed into bales, the juices of fruits extracted, and oils pressed from vegetable bodies, such as linseed and the almond.

In contrast with these uses of the screw, depending on the immense pressure it can exert, is another, remarkable for its delicacy. It is used to measure very small distances when accuracy is required. Screws with threads of exceeding fineness are used for this purpose. Suppose a screw with 100 threads in one inch of its length; then, at every turn its end would advance just $\frac{1}{100}$ of an inch, and if it carry a steel marker, spaces of that length may be marked off on any body along side of which it moves. Now let the power move in a circle 10 in. in circumference, and let this circle be graduated to inches, tenths, and hundredths. If the power move one inch on this scale, the marker on the end of the screw will go forward only $\frac{1}{1000}$ of an inch. If the power goes $\frac{1}{10}$ inch, then the marker will advance only $\frac{1}{10000}$ of an inch, a distance quite too small to be seen except by the aid of a good microscope. Astronomers use the micrometer screw to measure the apparent sizes of the heavenly bodies.

PROBLEMS ILLUSTRATING THE THEORY OF MACHINERY.

1. Suppose a body weighing 5 lbs. moves without resistance, with a velocity of 20 ft. a second: what momentum would it have? *Ans.* 100.

2. Suppose the body weighing 5 lbs. moves against resistance, with a velocity of 20 ft. a second, kept uniform by a constant force: what would be its living force? *Ans.* 2,000.

3. Suppose two bodies, one of 5 lbs., the other of 7 lbs. move, with equal velocities, 100 ft. a second, but in opposite directions, without resistance. Let them at the same instant strike a third body: in which direction would the body be moved, and with what force? *Ans. to 2d question*, 200 lbs.

4. Two trains of cars, each moving at the rate of 30 miles an hour, or 44 ft. a second, one weighing 10 tons, the other 20 tons, come in collision from opposite directions: what force would be exerted to dash them to pieces? *Ans.* 58,080 tons a second.

5. If a power of 10 lbs. act upon the long arm of a lever, a distance from the fulcrum of 6 ft.: what weight would it balance at a distance of 2 ft. on the other side of the fulcrum? *Ans.* 30 lbs.

6. In a lever of the second class, the power, 3 lbs., is at a distance of 1 ft. from the fulcrum: what weight will it balance at a distance of 1 in. from the fulcrum? *Ans.* 36 lbs.

7. In a compound lever, the long arms are 4 ft., 5 ft.,

and 6 ft. in length; the short arms are 1 ft., 2 ft., and 3 ft. long: a weight of 2,000 lbs. is to be balanced: how much power must act upon the first long arm?

Ans. 100 lbs.

8. A power of 10 lbs. lifts a weight of 500 lbs. by means of a lever whose short arm is 1 ft. long: how long is the long arm of the lever? *Ans.* 50 ft.

9. If the 500 lbs. in the last example is to be lifted 2 ft., how far must the power move to do it?

Ans. 100 ft.

10. The radius of a wheel is 30 inches; of its axle, 5 inches: a power of 100 ounces is exerted upon the wheel: how much weight will it balance at the axle?

Ans. 600 oz.

11. Three wheels and axles are combined, as shown in Fig. 50; the radius of each wheel is 20 inches; of each axle, is 4 inches; a power of 2 pounds acts on the first wheel: what weight will it balance on the last axle?

Ans. 250 lbs.

Fig. 60.

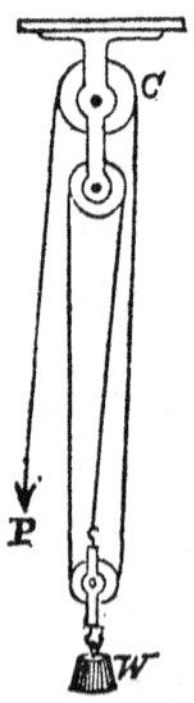

12. A force of 16 lbs. is applied to the last axle (Fig. 50), and moves at the rate of 10 inches a second: how much weight at the first wheel would balance it at rest? and how much slower will it go when in motion?

Ans. .128 lb.; $\frac{1}{12}$ as fast.

13. With a single movable pulley a stone weighing 350 lbs. is to be lifted: what power must be exerted?

Ans. 175+lbs.

14. With the single movable pulley, shown in Fig. 60, what power at P would balance a weight of 250 lbs. at W? *Ans.* $83\frac{1}{3}$ lbs.

15. If the weight W, is lifted by the power, how far would the power move to lift the weight 1 ft.?
Ans. 3 ft.

16. In a system of 4 movable pulleys, with a single rope, what power would be needed to balance a weight of 500 lbs? *Ans.* $62\frac{1}{2}$ lbs.

17. Suppose each of the 4 pulleys has a separate rope, what power would then be needed? *Ans.* $31\frac{1}{4}$ lbs.

18. An inclined plane, 6 ft. in length and 2 ft. high, is used to put a barrel of flour upon a cart. The barrel weighs 196 lbs.: how much force must a man exert, pushing parallel to the length of the plane?
Ans. $65\frac{1}{3}$+lbs.

19. If the base of the plane were 5 ft., its height 2 ft., and the man pushes parallel to the base, how much force must he exert to lift the barrel of flour?
Ans. $78\frac{2}{5}$+lbs.

20. The distance between the threads of a screw is 1 in., and the power of 25 lbs. moves in a circle of 3 ft. in circumference: how much weight will it balance?
Ans. 900 lbs.

21. A power of 20 lbs., by means of a screw, exerts a pressure of 800 lbs. The threads are $\frac{1}{2}$ in. apart: what is the circumference of the circle in which the power moves? *Ans.* 20 inches.

§ 5. OF THE MOTION OF LIQUIDS.

(37.) Water will issue from an opening in the side of a vessel with the same velocity which a body would gain by falling from the surface of the water to the center of the opening.

Hence the velocity of the jet of water, will depend only on the distance of the orifice below the surface of the water in the vessel, and may be calculated by the formula, $V = 2\sqrt{Sg}$.

1. *The velocity of a jet of water the same as that of a falling body.*—To prove this principle, we must remember: first, that water, confined in pipes, will rise as high as the source from which it comes [see (11.) 1]; second, that a body thrown upward, starts with the same velocity that it has when it gets back. (See p. 94.)

In Fig. 61, a bent tube (A) extends from near the bottom of a vessel of water. The water rises as high in the tube as in the vessel; it is the upward pressure of the water at A that pushes it up. The *same force* would be exerted on the water at A, if the tube were cut off at that point, and it would, if not resisted, throw the water to the *same height*, as shown on the other side of the figure, at B. But the velocity with which it must start from B to reach the level of R, is the same it would gain by falling from that level back to B. If the tube were cut off at C, the water would issue with the *same force*, and, therefore, with the same velocity. Hence *the velocity with which the water issues, is the same as that of a body falling from the surface of the water down to the center of the orifice.*

Fig. 61.

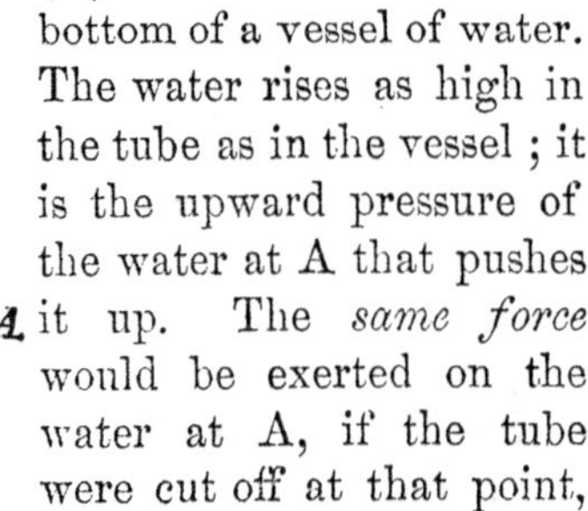

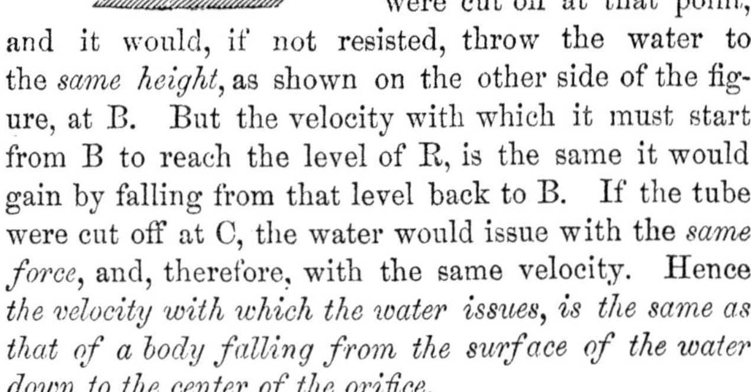

2. *The velocity of the jet depends on the distance of the orifice below the level of the water.*—The velocity of a falling body depends only on the height from which it has fallen. All bodies, whatever be their size

or nature, fall with equal velocities. In the same manner, all liquids, however different in nature, will issue with equal velocities, if the openings from which they are thrown are at the same distance from the surface of the liquid in the reservoir.

3. *Velocity calculated by the formula*, $V = 2\sqrt{Sg}$.—Now, the velocity of a falling body is given by the equation $V = 2\sqrt{Sg}$. [see (24,) 6], and it is clear that the velocity of a jet of water will be given by the same formula, if S represents the distance of the orifice below the level of the water in the vessel.

If, for example, we would know the velocity of a jet of water from an orifice 36 feet below the surface in a reservoir, we put 36 for S in the formula. It then reads $V = 2\sqrt{36 \times 16}$. The value of V is, then, 48; the velocity of the water is 48 feet a second.

(38.) The quantity of water discharged from an orifice, depends upon its velocity, the size of the orifice, and the time of flow. It may be found by multiplying the values of these three things together.

1. *To calculate the quantity.*—If the orifice have an area of 1 sq. ft., then the *velocity* will represent the number of *cubic feet* discharged in one second. Multiplying this by the number of square feet, or fraction of a square foot in the orifice, must show the number of cubic feet flowing from the orifice in one second, and this multiplied by the number of seconds, will tell the number of cubic feet discharged. For example, how much water will flow from an orifice of $1\frac{1}{2}$ sq. ft. area, at a depth of 9 ft. below the surface of the water, in 10 seconds? At a depth of 9 ft. the water will issue with

a velocity, $V = 2\sqrt{9 \times 16} = 24$ ft. Now, if the opening was *one square foot*, then 24 cubic ft. would issue in *one* second, and $24 \times 1\frac{1}{2} \times 10 = 360$ cubic ft. must issue from the orifice of $1\frac{1}{2}$ sq. ft. in the given time, 10 seconds.

The rule is concisely expressed by the formula:—

$$Q = V \times A \times T, \text{ in which}$$

Q represents the quantity of water discharged,
V " " Velocity,
A " " Area of the orifice,
T " " Time of flow.

In this equation there are four things, and it is clear that, any three of them being given, the fourth, whichever it may be, can be found. A single illustration will show how this is done.

Suppose 10,000 cubic ft. of water must be discharged in 60 seconds, from an orifice so far below the surface of the water that the velocity of the jet is 250 ft. a second: how large must the orifice be made?

In this problem, the value of V is given, 250 ft.; the value of T is 60 seconds; the value of Q is 10,000 cubic ft.; the value of A is wanted. By putting the given values into the equation it becomes:—

$$10{,}000 = 250 \times A \times 60; \text{ hence,}$$
$$A = \tfrac{2}{3} \text{ sq. ft., or 96 sq. in.}$$

(39.) The velocity of a jet of water and the quantity discharged are found in practice to be much less than the foregoing theory would give. The actual amount may be increased by using short tubes of different shapes.

1. *The velocity in practice less than in theory.*— If the experiment be tried with a vessel of water, as

shown by Fig. 61, it will be seen that the jet does not rise quite as high as the level of the water in the vessel. It does not, because the resistance of the air prevents it. From any orifice, water must issue against the resistance of air, and its motion is less rapid on that account.

2. *The quantity in practice less than in theory.*—If we examine a jet of water flowing from an orifice in the side of a vessel, we will see that it grows rapidly smaller, so that, at a little distance, its size is only about $\frac{2}{3}$ as great as at the orifice. Beyond this point the contraction of the jet is gradual. The rapid contraction near the orifice is due to *cross currents*, caused by the water flowing toward the orifice from different directions in the vessel; these currents may be seen if there be any solid particles floating in the water. If the jet were the size of the orifice, the quantity of water discharged would be what the theory gives, but since it is only about two-thirds as large, there will be only about two-thirds as much water discharged.

3. *The quantity increased by using tubes.*—Short tubes inserted in the orifice are found to increase the actual flow. These tubes are either *cylindrical or conical.*

It is found that a cylindrical tube, whose length is not more than four times its diameter, if placed in the orifice, will increase the amount discharged to about .82 of that which theory gives. In this case the water *adheres* to the sides of the tube, so that the contraction of the jet is prevented; the jet is the size of the orifice. By the use of conical tubes the amount discharged may be made still greater. (See Silliman's Physics, pp. 174. and 180).

(40.) Water-wheels are turned by the power of moving water. There are several kinds: First, the undershot wheel; second, the overshot wheel; third, the breast wheel; fourth, the turbine wheel.

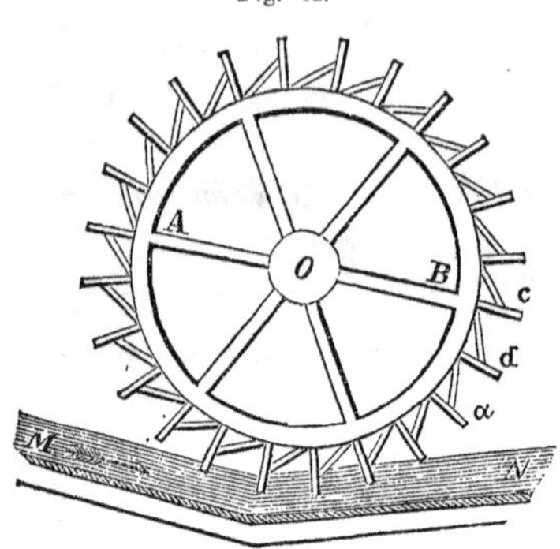

Fig. 62.

1. *The undershot wheel.*—The undershot wheel is shown in Fig. 62. Its circumference is provided with *float-boards a b c*, against which the running water acts. Other wheels are connected with the axle of this one by cogs and bands. This form of wheel is often placed in a horizontal position, and water from the bottom of a dam guided against the float-boards of one side.

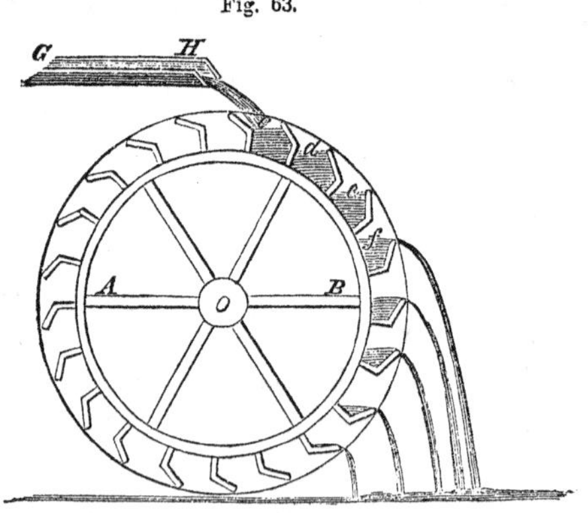

Fig. 63.

2. *The overshot wheel.*—The overshot wheel (Fig. 63)

differs from the undershot, by having buckets upon its circumference, instead of float-boards. The water enters the buckets at the top of the wheel, and, filling those on one side of it, turns the wheel by its weight. The buckets all open in the same direction, so that while those on one side of the wheel are full, those on the other side will be bottom upward and empty.

3. *The breast wheel.*—The breast wheel (Fig. 64) differs from the undershot wheel only by being so

Fig. 64.

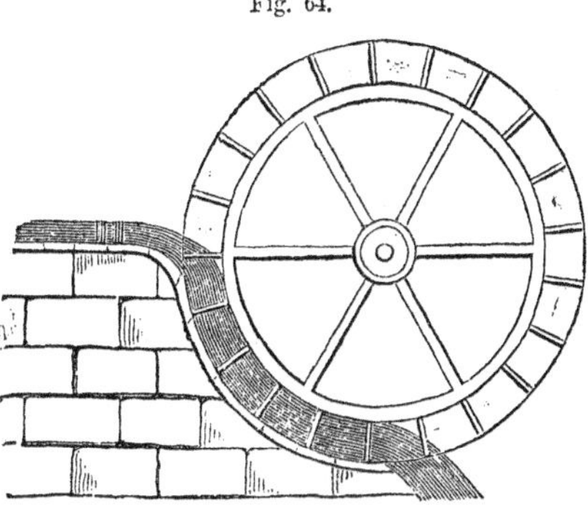

placed in front of a dam, that the water shall fall upon the float-boards of its circumference on a level with its axis.

4. *The American turbine.*—The construction of the turbine is more complex than the wheels just described. Its action may be understood by a careful study of Fig. 65.

This figure shows a section of the *interior* of the wheel, as it would appear to one who looks down upon it as it lies in its horizontal position. In the center is a circular disk of cast iron, A B, in a horizontal position. On the upper surface of this disk are fastened the curved guides, *a a a*. This disk *is stationary*. The wheel proper, C D, revolves outside of this disk. It consists of two cast-iron plates, one above the other, the space between them being divided into numerous channels by the curved partitions, *c c c*. The partitions in the wheel, and the guides on the disk, are curved in opposite directions. To the under plate of this wheel is fastened a cast iron plate, which extends *under* the central disk A B, and to the center of this plate is attached a vertical shaft which comes up through the disk at E.

Fig. 65.

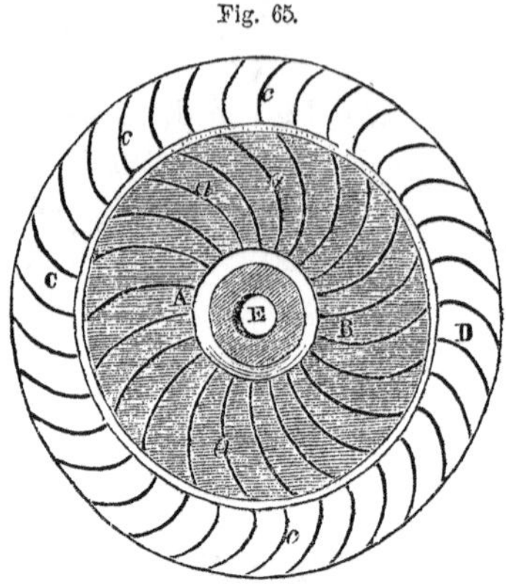

The turbine, except the upper part of the shaft, is *entirely under water*. The weight of the column of water above the disk forces the water with great power and force out from between the curved guides of the disk, into the curved channels of the wheel, in as many different streams as there are spaces between the guides. *The force of these streams*, striking against the partitions of the wheel, turns the wheel in the direction indicated by the arrow. The vertical shaft turns with the wheel, and, by means of cogs, gives motion to other parts of the machinery. (See Silliman's Physics, p. 184.)

Of all forms of water-wheel, the turbine is most energetic and economical.

§ 6. OF THE MOTION OF AIR.

(41.) Air in motion is called wind. Winds are produced by the action of heat and the attraction of gravitation upon the atmosphere; and, in the case of the trade winds, partly by the rotation of the earth on its axis.

1. *Wind.*—The motion of air, called wind, is due to a difference in the temperature of two portions of the atmosphere. Heat expands air. One hundred cubic inches of hot air will *weigh less* than a hundred cubic inches of cold air. Now, if a portion of hot and light air is surrounded by that which is colder and heavier, it will rise, for the same reason that a cork rises in water. It will be pushed up out of the way by the heavier air, which takes its place.

Let us now suppose that, in some particular part of the country, the air becomes heated more than in surrounding portions. This heated and lighter air will be pushed up by air *moving in* from all directions to take its place. This moving air is wind. People residing north of the heated place will observe a north wind, and those south of it a south wind.

Now, there is an unequal distribution of heat over the surface of the earth. It is caused partly by the changes of the seasons, and partly by various local causes. To it the *production* of winds is due. Their direction will be modified by many causes: the form of the surface over which they pass is an important one. As the same wind often blows in different directions on different sides of a house; or, as blocks of buildings compel the wind to sweep up and down the various streets of a city, so the hills and valleys of a country, or the presence of forests or plains, will modify the direction of the winds that blow over them.

2. *The trade winds.*—The trade winds require particular notice. They occur in the equatorial parts of the earth, *and always blow in the same directions.* Over a surface of about 30° of latitude on the north side of the equator, they blow from the northeast toward the southwest; while south of the equator, over about the same width of zone, they blow from the southeast toward the northwest. These directions are maintained so constantly, that mariners count upon the trade winds with almost the same certainty as upon the rising and setting of the sun.

3. *Due to heat and the rotation of the earth.*—To explain this phenomenon we must remember: first, that the equatorial region is constantly heated by the sun more than parts of the earth either north or south; and second, that the earth revolves from west to east, the equatorial parts moving most swiftly.

The heated air at the equator, lighter than the air either north or south of it, will be pushed up, while currents of colder air from the north and from the south, will move toward the equator. But the equatorial parts of the earth move toward the east more swiftly than other parts; the air from the north must, therefore, pass over portions of the earth which move eastward faster than itself, and it will be left behind. We find, then, that there is a real motion from the north, and at the same time an apparent motion from the east; these two motions combined make the direction of the wind to be from the northeast. A similar explanation will show why the southern trade wind blows from the southeast toward the northwest. (See Silliman's Physics, p. 643.)

CHAPTER IV.

OF MOTION—(CONTINUED).

INTRODUCTION.—APPLICATION OF THE FUNDAMENTAL IDEAS.

(42.) Read (4), (7), and (20).—Attraction, repulsion, and inertia, acting upon masses, or molecules, produce vibrations.

1. *Attraction, repulsion, and inertia.*—We have seen that the forces of nature are only different manifestations of attraction and repulsion. [See (20.) 1.] We have also seen that a body in motion can not stop itself. [See (4.) 2.] When, therefore, a body has been put in motion, by any force, it will move in that direction, on account of its inertia, until stopped by an opposite force. Suppose the force which stops it continues its action afterward, it will move the body back toward its first position, and then if the force cease, the inertia will move it onward until again stopped by an opposite force.

2. *Vibrations.*—The body, thus acted upon, will swing *alternately back and forth over the same path.* Such a motion is called *vibration.*

If, with the finger, we sink one scale-pan of a balance.

it will continue to pass alternately up and down over the same path for a long time after the finger is removed: it *vibrates.* Or if, instead of pushing it down we pull the scale-pan to one side and then let go of it, it will swing back and forth for a long time; this alternate motion, to and fro, is *vibration.* Suppose a ball, hung by a fine wire, be twirled by the fingers so as to twist the wire; let go of it, and, speedily untwisting the wire, it will go on for a time twisting it up the other way. The ball rotates, first in one direction and then in the other, and this alternate motion is *vibration.*

Or take a bent glass tube; pour water into it until the arms are two-thirds full; tip it to one side and then suddenly bring it back to a vertical position. The water will rise and fall in the arms of the tube, and will continue this alternate motion up and down for some time. In this case a *liquid* vibrates.

Gases may be made to vibrate in the same way.

§ 1. OF THE VIBRATIONS OF THE PENDULUM.

(43.) The pendulum vibrates under the influence of gravitation and inertia. Its vibration is governed by three laws:—

1st. The time of one vibration varies as the square root of the length of the pendulum.

2d. The time of one vibration varies inversely as the square root of the force of gravity.

3d. The time of one vibration is independent of the length of the arc through which the pendulum vibrates.

1. *The pendulum.*—A body hanging from a fixed point by a flexible cord or wire, is called a *pendulum*

In Fig. 66, the pendulum is represented as a ball B, hung from a point A.

If this ball be lifted from the point B to C, and then loosed from the hand, it will swing back and forth through the arc D C, going a less and less distance, until finally it will stop at B. Its motion, from one end of its arc D, to the other C, is a single *vibration*, and the *distance* D C, through which it vibrates is called the *amplitude* of vibration.

Fig. 66.

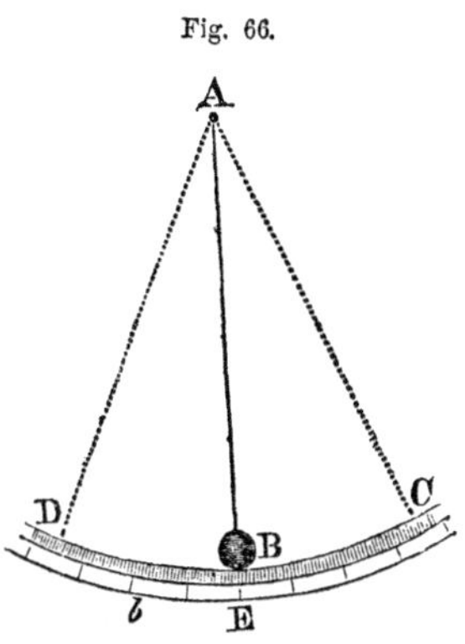

2. *It vibrates under the influence of gravitation and inertia.*—Suppose a ball at M (Fig. 67), to represent a pendulum hung from the fixed point C, by a cord M C. Now, if this ball be lifted to the point *m*, and, for a moment, held there, the force of gravity will act upon it in a vertical direction. We will represent this force by the line *m* A, and resolve it into two components [see (26.) 1], shown by the lines *m* D and *m* B. The force, *m* B, acts lengthwise of the string *without effect to move the ball:* the other force, *m* D, at right angles to the first, will pull the ball toward the point M. If the ball is allowed to fall to M, its inertia will carry it beyond that point; but gravitation will then be pulling it back with just the *same power* that it exerted to pull the ball from *m* to M. It will rise from M to *n*, a distance just as far

Fig. 67.

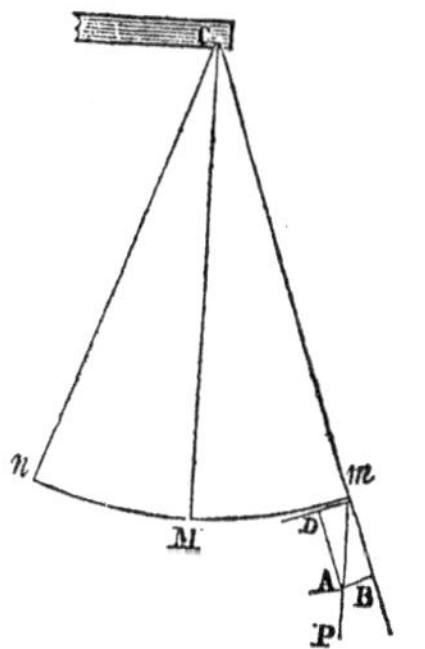

from M, as it has fallen from *m*. It will there stop, and gravitation will bring it back to M, while its inertia will carry it up to *m*, and if there were no resistance to its motion it would vibrate forever through the arc *n m*. The resistance of the air and the friction of the cord on the hook will finally make it stop at M.

3. *The first law.*—If two pendulums of different lengths (P and P', Fig. 68), be made to vibrate together, the short one will be seen to vibrate much faster than the other. We learn from this that the time of vibration depends on the *length* of the pendulum.

Fig. 68.

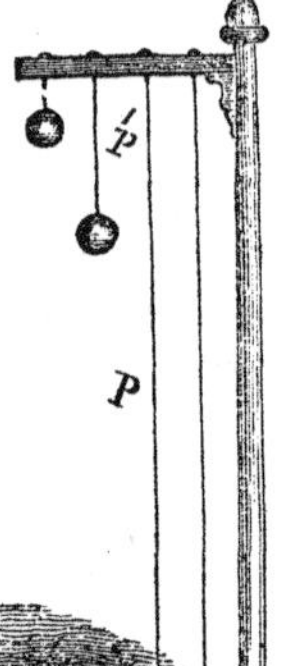

Now, let us make the pendulum P *just four times* as long as the other. With a watch in the hand, we can easily count the number of vibrations it makes in one minute, and 60 divided by this number, shows how long it takes to make *one* vibration. In the same way we can find the time it takes the shorter pendulum to make *one* vibration. Doing this, we find that the pendulum P, takes *twice* as long as the other to vibrate once. Being four times as long as the other, the time of vibration is two times as great. Hence, *the time of one vibration varies as the square root of the length of the pendulum.*

The length of a pendulum to vibrate in one second is about 39.1 inches; to vibrate in two seconds, it must be *four times* as long; to vibrate in one-half a second it must be *one-fourth* as long.

4. *The second law.*—By calculating the force of

gravity (see prob. 21, p. 94), at different distances above the level of the sea, and then, by experiment, finding the time of one vibration made by the *same* pendulum at those places, it will be found that *the time of one vibration varies inversely as the square root of the force of gravity.*

5. *The third law.*—Finally, if we make the pendulum P, vibrate in a large arc, and find the time of one vibration, and then make it vibrate in a small arc, we shall find the time of one vibration to be the same. The pendulum must vibrate in equal times, no matter whether its arc be large or small. In other words, *the time of one vibration is independent of the arc through which the pendulum vibrates.*

This third law is absolutely true only when the arcs compared are *very small.* Yet, in the latitude of Paris, it is found that for a pendulum whose length is one meter, or 39.37 in., the time of one vibration, through an arc of 8°, is only .000076 of a second longer than if its arc were infinitely small. (See Cooke's Chem. Phys. p. 69.)

(44.) These laws apply to a single point in a pendulum, called the center of oscillation.

1. *The center of oscillation.*—The different molecules of a pendulum are at different distances from the point of suspension, and hence would vibrate in different times if they were not held together by cohesion. Although they are held together, and must all move at once, yet the forces that would make them vibrate differently are acting just the same as if they were not. The upper parts of the pendulum are *trying* to vibrate faster, and must be pulling the lower parts along; while

the lower parts are *trying* to vibrate slower, and must be pulling the upper parts back. There must be some point in the pendulum, at which these two struggles just balance each other. This point will vibrate just as fast as if it were influenced by no other molecules whatever. *A point in the pendulum which vibrates as if only under the influence of its own gravitation and inertia* is called the *center of oscillation.* The center of oscillation is *generally* a little below the center of gravity of the pendulum ball.

2. *The laws apply to this point.*—The three laws, obtained in the foregoing paragraph, apply to only this point, the center of oscillation. Indeed, whenever we speak of the pendulum we refer to this point. By the length of a pendulum we mean the distance from the point of support to the *center of oscillation*, and when we use the term vibration, we refer to the motion of this one point of the pendulum.

(45.) There are several uses of the pendulum; we notice only two:—

1st. It is used to measure time.

2d. It is used to determine the form of the earth.

1. *Used to measure time.*—The vibrations of a pendulum are made in equal times. If then we know the time of one vibration, and can count the number made, we know the time during which the pendulum vibrates.

Now, the common clock is an instrument in which, by weights, friction and the resistance of air are overcome, so that the pendulum shall continue its motion, and by which, the number of vibrations are at the same time recorded by the hands moving over a graduated dial.

2. *Used to determine the form of the earth.*—The pendulum has been used to determine the shape of the earth. For this purpose, pendulums of the same length have been made to vibrate in different latitudes. It has been found that the time of one vibration is less and less as the pendulum approaches the poles of the earth. Now, to make the vibrations more rapid, the force of gravity must increase, and if this force is stronger toward the poles, the surface of the earth must be nearer the center of the earth there than at the equator. The polar diameter must, therefore, be shorter than the equatorial diameter, and the shape of the earth must be that of an *oblate spheroid.*

§ 2. OF THE VIBRATIONS OF CORDS.

(46.) The vibrations of cords are due to the action of elasticity and inertia. They are governed by three laws:—

1st. The number of vibrations in a second varies inversely as the length of the cord.

2d. The number of vibrations in a second varies directly as the square root of the weight by which the cord is stretched, or its tension.

3d. The number of vibrations in a second varies inversely as the square root of the weight of a given length of the cord.

1. *The vibration of cords.*—Let a cord or string be stretched between two fixed points (*a* and *b*, Fig. 69).

Fig. 69.

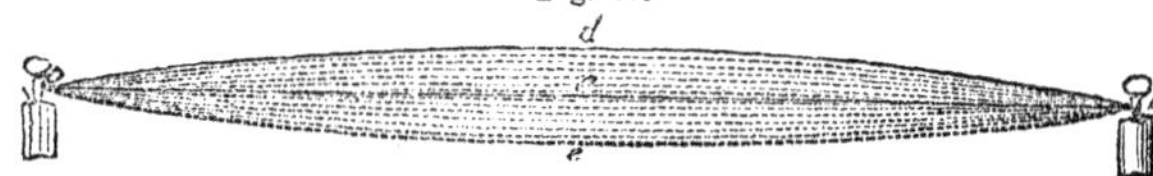

By taking hold of its middle point, the cord may be

drawn to one side, *a e b*. Then loose it, and it will spring back and go an equal distance on the other side *a d b*; then return, and so continue to swing rapidly back and forth until it finally stops in its first position, *a c b*.

The motion of the cord from *e* to *d* and back again, is a complete *vibration*. Its motion from *e* to *d*, is a half vibration, or, as generally called, a single vibration. The distance from *e* to *d*, is the amplitude of vibration.

2. *Due to elasticity and inertia.*—When the force which stretches the string into the position *a e b*, is withdrawn, elasticity moves it back to its first position, *a c b*, and the inertia gained by this motion, throws it forward an equal distance, to *a d b*. The elasticity of the string again pulls it back to the position, *a c b*, and its inertia carries it beyond, and thus, under the joint influence of elasticity and inertia, the string will swiftly vibrate, its amplitude growing less and less, on account of resistance, until at last it stops in its first position.

3. *The laws of vibration.*—The vibrations of cords are, in all cases, quite too rapid to be counted, and yet it will be impossible to establish any laws of vibration, unless we can find the *number* of vibrations made in a given time. How can this be done? *

However rapid the motion of the cord may be, the lightning swiftness of electricity is yet greater; so the cord, by using electricity, may register the vibrations which it makes.

The apparatus used for this purpose by the author, is shown in Fig. 70, as far as necessary to illustrate the principle of the process.

* The syren will be described in the chapter on sound: it seems desirable here to make the cord directly register its own vibrations, so that the *laws of vibration* shall stand independent of sound.

A cord, A B, rests upon the two bridges, C and D, and passing over a pulley, B, is stretched by a weight, W. Through its middle point is a fine cambric needle *n*, just under the point of which stands a vessel of mer-

Fig. 70.

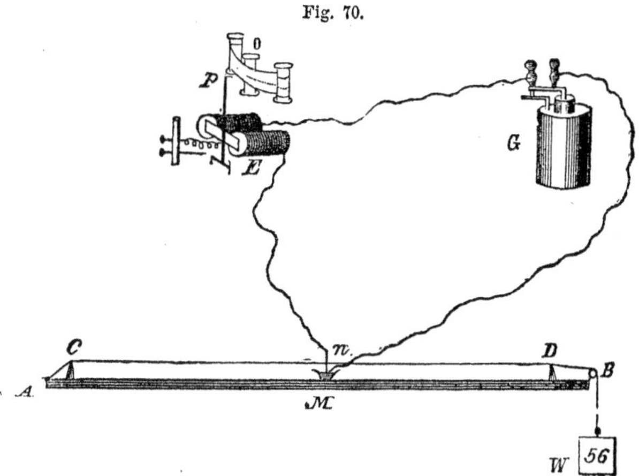

cury, M. From an electrical battery, G, one wire goes to the mercury, and another, after passing around an electro-magnet, E, is threaded into the eye of the needle. At P, is a sharp and soft pencil-point, and in front of it is a roller O, over which passes a strip of paper.

If now, the string vibrates up and down, the point of the needle will come in contact with the mercury below it at the end of every vibration. When the needle touches the mercury the electricity darts through the wires, and the magnet E, instantly pulls the pencil-point against the paper, and a *dot* is thereby made.

If the paper be drawn along in front of the pencil-point while the string is vibrating, a series of dots will be made, and the number of dots shows the number of vibrations made by the string.

The apparatus by which the paper is drawn along is not shown in the figure, neither is that by which time is measured.

With this apparatus we proceed rapidly to verify the laws of vibration.

4. *The first law.*—The string C D, was taken 3 ft. in length: stretched by a weight of 56 lbs. at W, it made 420 complete or double vibrations in 3 seconds. The bridges, C and D, were then moved, so that the length of the string was 4 ft.; it then made 315 vibrations in 3 seconds. But 420 is to 315 as 4 : 3. We see that when the lengths of the string are as 3 : 4, the number of vibrations in the same time are as 4 : 3. Hence the number of vibrations in a given time varies inversely as the lengths of the string.

5. *The second law.*—The string was again made 4 ft. long, and the weight W, 56 lbs. The vibrations in one second then numbered 105. When the weight, W, was then changed to 14 lbs., the number of vibrations in one second was, in some experiments 52, and in others 53. The instrument can not register parts of a vibration; the true number is evidently between 52 and 53; we may call it 52½. We see that when the weights are 56 and 14, or as 4 : 1, the number of vibrations made in a second are 105 and 52½, or as 2 : 1. Hence the number of vibrations in a second, varies directly as the square root of the weight by which the string is stretched.

6. *The third law.*—The string which, being 4 ft. long, and stretched with a weight of 56 lbs., gave 105 vibrations a second, was found to weigh 19.4 grs. to the foot in length. Another string, weighing 43 grs. to the foot, was taken of the same length and tension as the other, and the number of vibrations in one second was, in

some experiments 70, and in others 71. The true number is between these; call it $70\frac{1}{2}$. Now, the weight of equal lengths of the string being 19.4 : 43, the number of vibrations are found to be $105 : 70\frac{1}{2}$; but $105 : 70\frac{1}{2} :: \sqrt{43} : \sqrt{19.4}$, so nearly that we may infer, that the number of vibrations a second varies inversely as the square root of the weights of equal lengths of the string. [See (107.) 6.]

Fig. 71.

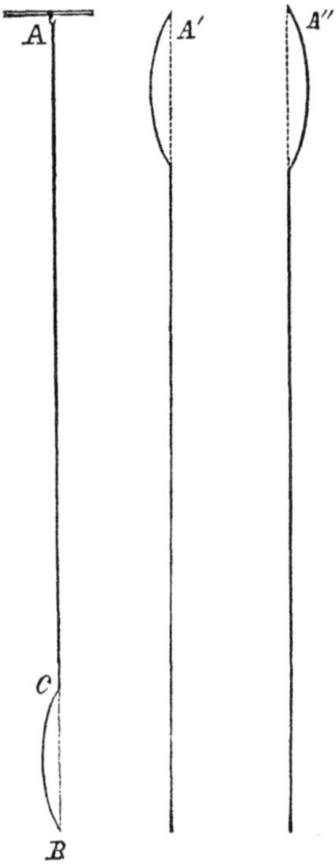

(47.) In progressive vibrations, the motion appears to be lengthwise of the cord. A cord may divide itself into parts, vibrating separately, called ventral segments, with points of rest between them, called nodes.

1. *Progressive vibrations.*—Let a heavy cord, or better still, an india-rubber tube (A B, Fig. 71), several feet long, be fastened at one end to the wall or ceiling of the room. Take hold of the other end with one hand, and by a sudden blow with the other, push the part B C aside, as shown in the figure. The little hillock thus formed will run swiftly up the tube to A, and then quickly down to the hand again. By carefully noticing the motion, it will be seen that while the hillock, running up to A,

is on one side of the cord or tube, that which returns to the hand is on the other. Having gone to the top, as seen at A′, it turns as seen at A″, and then comes down. Nor does it then stop; it will again and again run up and down the tube until, the height of the hillock growing less and less, it finally disappears. This motion is progressive vibration.

Fig. 72.

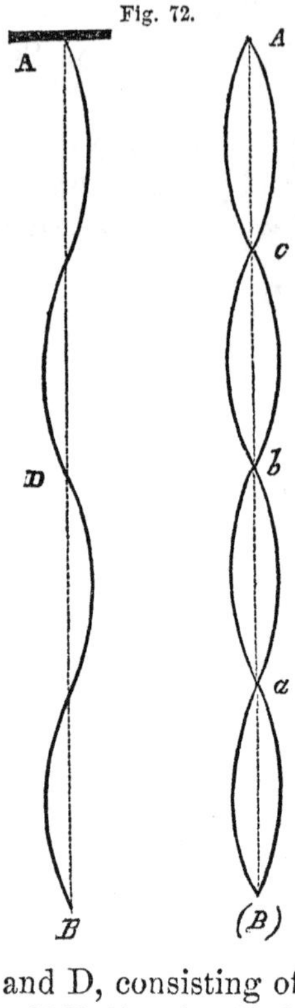

2. *The motion appears to be lengthwise of the tube.*—It is interesting and important to notice that while the motion appears to be lengthwise of the cord or tube, the only *real* motion of the parts, is back and forth, across their first position.

3. *Ventral segments.*—By starting several hillocks, one after the other quickly, the whole cord may be thrown into a series of hills and valleys, as shown in Fig. 72. In this case the motion between B and D, consisting of two parts, on opposite sides of the middle line, is called a *wave.* Two waves are represented in the figure.

By skillfully timing the impulses of the hand, the hillocks on both sides of the middle line in Fig. 72, may be made to turn themselves over at the same time. In that case, the tube will present the appearance shown in Fig. 72 (B), the points *a b c*, being *almost* stationary while the parts between are swinging to and fro across the middle line, making vibrations, just as if they were separate cords.

The points which appear to be at rest are called *nodes*, while the vibrating parts between them, are called *ventral segments.*

(48.) When a cord fastened at both ends is struck, it vibrates as a whole, and in ventral segments at the same time.

1. *It vibrates as a whole.*—Suppose the cord shown in Fig. 69, to be struck at a point one-sixth of its length from the end *b*, the entire cord will swing back and forth just as represented in that figure, and the vibration of its whole length will be governed by the three laws already given, in (46).

2. *Ventral segments at the same time.*—The cord will at the *same time* divide itself into *three* ventral segments, each of which will make a series of separate vibrations, while taking part in the full length vibration of the cord.

Now, as each segment in this case is $\frac{1}{3}$ the length of the string, it must (first law) vibrate 3 times as fast. If the string is struck at $\frac{1}{4}$ of its length from the end, there will be 2 segments, each $\frac{1}{2}$ as long as the string, and of course, vibrating 2 times as fast.

§ 3. OF VIBRATIONS IN LIQUIDS AND GASES.

(49.) Progressive vibrations are illustrated by water waves.

Two sets of water waves may interfere with each other, and produce a single set different from either.

1. *Water waves.*—Let a pebble be tossed into the

water of a lake or pond, and the tranquil surface will be carved into a series of circular ridges and furrows, which, growing gradually larger and larger, finally break against the shore. The motion appears to be in all directions outward from the pebble, but the little sticks and straws that may be resting upon the water at the time, tell us, by their dancing, that the *real* motion of the water is, like their own, a motion only up and down.

A *wave* of water consists of two parts, a ridge and a furrow.

2. *Water waves may interfere.*—Let two sets of water waves be started at the same time, by dropping two pebbles at a little distance from each other. The two sets of growing circles very soon cross each other, and then the smooth surface of the water will be cut up into a curious confusion of dancing hummocks. Some of these hummocks will be twice as high as the ridges of either set of waves, while others will just lift their heads above the original surface of the water. When two sets of waves are thrown together, they are said to interfere.

But why are the hummocks of such different heights? It is clear that when two ridges come together their heights will be united, and the height of the hummock will be the *sum* of their separate heights. But when the ridges of one set enter the furrows of the other, the height of the resulting hummock will be equal to their *difference*. Now, as the waves are running across each other, the hummocks must be of various heights, limited on the one hand by the *sum*, and on the other, by the *difference* in the heights of the ridges of the two sets.

(50.) The vibrations of air consist of alternate rarefactions and condensations. In free air the waves travel outward from their source in every possible direction. Different sets must be constantly interfering.

1. *Alternate rarefactions and condensations.*—We have seen [see (16.) 1 and 2] how easily air may be compressed, and with what promptness it springs back to its former volume when the compressing force is removed. Now, suppose that near to one end of a long tube, is a piston P (Fig. 73). By suddenly pushing this piston forward to P′, and then instantly pulling it back, the air in the whole length of the tube will be put in motion. Let us analyze this motion.

Fig. 73.

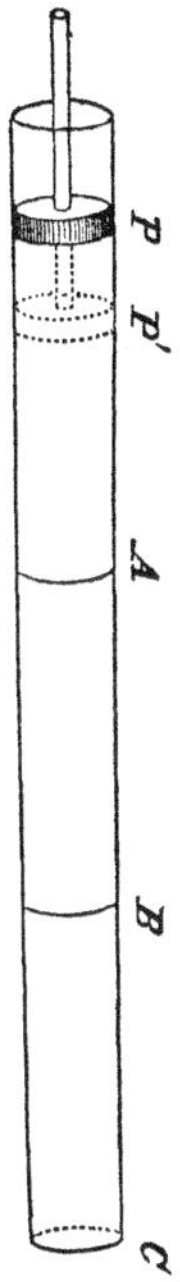

When the piston moves from P, it crowds the air before it, and when it has reached P′, this crowding effect will have gone forward to some point A, more or less distant. The space, P′ A, is then filled with condensed air. Now, when the pressure of the piston is removed, the condensed air springs back. It springs both ways, backward against the piston and forward against the air at A. By its pressure against the air at A, the air in the space A B, will be condensed. The next moment this air expands, and pressing both ways, condenses the air B C, in front of it and also the air A P, behind it. These two portions will, in this way, be condensed, while the air, A B, will be rarefied. The next instant these condensed portions spring back and become rarefied, while the rarefied portion A B, and at the same time, another part beyond C, will be condensed. The air is in this way thrown into a series of condensed and rarefied parts, alternately springing back and forth in the direction lengthwise of the tube. We need only add, that there is no *sudden* transition from condensed to rarefied air at the points A B and C. The mobility of air will not permit this. At the middle of the condensed part the condensation is greatest, while at the middle of the rarefied part is the greatest rarefaction, and between these points the change is gradual.

A *wave* of air consists of two parts, a condensation and a rarefaction.

2. *Waves in free air go in all directions.*—The walls of the tube confine the air, and compel its waves to go in the direction of its length; in free air the case is different. Every impulse by which the atmosphere at any point is suddenly condensed or rarefied, is the center from which air waves go outward in all directions.

Let a few grains of gunpowder be exploded. A little *sphere* of air at the point where the explosion occurs, will be, for the moment, rarefied, while by its pressure a *shell* of air outside of it will be condensed. This condensed air instantly springing back, condenses the air on both sides of it, and itself becomes rarefied. The waves will thus travel outward from the center, until the whole body of air is thrown into a series of concentric *shells*, alternately condensed and rarefied.

How constant and complicated must be these vibrations of the air! Every sudden and local puff of wind; every forcible breath exhaled from the lungs; the fall of every stick and stone, all these are the sources of as many different sets of waves spreading in all directions, darting across and through each other, too delicate to be seen or felt, presenting to the mind a scene of activity far exceeding the power of the senses to appreciate.

3. *Different sets interfere.*—Suppose two sets of air waves come together; if their condensed parts coincide, a single set will be formed whose condensations are greater than either. If the condensed parts of one set coincide with the rarefied parts of the other, there will be a single set whose condensations are less than either. In the first case, if the two sets are equal, the resulting waves will be doubled; if, in the other case, the two sets are equal, they will destroy each other, leaving the air without waves.

§ 4. OF THE VIBRATIONS OF MOLECULES.

(51.) The molecules of all bodies are at all times in motion. These vibrations of molecules can not be seen, yet they are able to affect our senses. Acting upon the ear they produce *sound;* upon the eye they are recognized as *light;* while upon the sense of touch they produce *heat.*

1. *Molecules in motion.*—The molecules of bodies do not touch each other; if they did, they could never be pushed nearer together, and there could be no such thing as elasticity. They are distinct, separate bodies. Moreover, they are supposed to be *in rapid motion.* Just how they move is not known. They may be swinging back and forth in straight lines, or in curves; they may be rolling on their axes to and fro, or perhaps revolving around each other: or it may be that they make several of these motions at once. Be this as it may, they are supposed to be in *motion* of some kind.

The vibrations of the molecules may be increased or diminished. To illustrate: look upon a bar of iron; imagine the multitude of little molecules of which it is made; see them in rapid vibration, trembling in their little spaces. Now strike the bar with a hammer; the hammer can not stop without giving the force which moves it, to the molecules of the bar, and every molecule acted on by this force, has its vibrations thereby quickened.

2. *These vibrations affect the senses.*—The motions of the molecules are quite too delicate to be seen. They are supposed to exist, only because many effects can be explained in no other way so well as on this supposition. They are thought to be the means by which objects of matter produce effects on our senses. The organs of sense are so arranged by Him who made them, that each one receives a different effect, although the vibrations that produce it may in all cases be much alike. The ear is so made that vibrations in it produce sound. The eye is so made that vibrations are recognized as light. The sense of touch is so arranged that vibrations against it are felt as heat. The phenomena of sound and light and heat are caused by vibrations. How simple the means to produce such wonderful results! "Know ye, that the Lord he is God; it is he that hath made us, and not we ourselves!"

CHAPTER V.

THE EFFECTS OF VIBRATIONS.—I. SOUND.

§ 1. THE ORIGIN AND THE TRANSMISSION OF SOUND.

(52.) READ (51). Sound is a sensation produced in the ear by the vibrations of external bodies.

1. *Sound produced by vibrations.*—Let two books be clapped together, and every ear in the room receives a shock, to which the name of sound is given. The molecules of the books are made to vibrate by the blow, and these vibrations, acting upon the air in contact with them, produce air waves. These air waves, traveling outward in all directions, finally reach the ear, and the many parts of this organ receiving these vibrations, enable the mind to recognize the peculiar sensation which we call sound.

When we listen to the sound of a church bell, we may in like manner imagine the molecules of the bell all in a state of tremulous motion, caused by the blows of the hammer. This motion causes vibration in the air in contact with the bell. The air waves thus formed, travel in all directions from the bell until the ear receives them.

The roar of a cataract is the result of vibrations caused by the falling water. The rolling sound of thunder is the effect of vibrations in air, caused by electricity. Every sound in nature, or that can be produced by art, may be traced back through the waves of some medium, to the vibrating molecules of some solid, liquid, or gaseous body.

(53.) Sound waves travel through all elastic media or bodies. The velocity of sound is not the same in different substances; it is governed by two laws:—

1st. The velocity of sound varies inversely as the square root of the density of the substance.

2d. The velocity of sound varies directly as the square root of the elasticity of the substance.

In the same medium, the velocity of sound is uniform.

1. *Sound waves.*—All sounds are the effects of vibrations, but it is not true that all vibrations produce sound. There are vibrations too slow to affect the ear; such are the vibrations of a cord not over-stretched. On the other hand, there are vibrations which are too rapid to be heard. The lower limit has been fixed at 16 vibrations a second, and the higher at 38,000. Waves which occur within these limits of velocity can be heard, and are called *sound waves.*

It is interesting to notice that the limits of hearing are not the same in all persons. "Nothing can be more surprising than to see two persons, neither of them deaf, the one complaining of the penetrating shrillness of a sound, while the other maintains that there is no sound at all." "In the 'Glaciers of the Alps' I have referred to a case of short auditory range noticed by myself in crossing the Wengern Alp in company with a friend. The grass at each side of the path swarmed with insects which, to me, rent the air with their shrill chirruping. My friend heard nothing of this, the insect music lying quite beyond his range of audition." (See Tyndall's Lect. on Sound.)

2. *Are transmitted through all elastic bodies.*—Numerous facts easily verified, prove this statement. When, for example, the blows of a hammer fall upon one end of a long wooden beam, an ear placed in contact with the other end hears the sound with surprising distinctness. The same thing is true of other solid bodies. The clatter of horses' hoofs, or the rattle of a railway train, quite inaudible to one who stands erect, is heard distinctly when the ear is placed in contact with the ground. The solid earth transmits the sound waves.

In liquids, also, sound waves travel freely. Let two stones be struck together under water; the sound will be heard by an ear, itself under water, a long distance away.

The transmission of sound waves through gases is sufficiently familiar; the sounds which throng the ear so constantly are transmitted through the atmosphere.

3. *The velocity not the same in all media.*—The velocity of sound in a great many substances, has been found by laborious and skillful experiments (see Tyndall's Lect. on Sound, p. 26). In the following table some of these results are collected :—

SUBSTANCES.	TEMPERATURE.	VELOCITY.
Air....	32° F.	1,092 ft.
Air.....	61 "	1,118 "
Oxygen..........................	32 "	1,040 "
Hydrogen.......................	32 "	4,164 "
River Water.....................	59 "	4,714 "
Iron............................	68 "	16,822 "
Pine Wood......................	..	10,900 "

The velocity of sound depends upon the *density* and the *elasticity* of the medium in which it travels.

4. *The first law.*—The density of oxygen, other things being equal, is about 16 times that of hydrogen. But we see in the table that the velocity of sound in oxygen, is only about $\frac{1}{4}$ as great as in hydrogen. In this case the velocity is inversely as the square root of the density of the medium. This law may be verified by repeated experiments.

5. *The second law.*—When air is heated in a tight vessel its elasticity is increased, while its density is unchanged. In this condition it will conduct sound more rapidly. If the elasticity of air be made 4 times as great, the velocity of sound will be doubled. The velocity of sound in this case is directly as the square root of the elasticity of the medium. It is so in all cases.

It is evident that both density and elasticity must be known, before we can judge the power of a substance to conduct sound. Liquids are, for example, more dense than gases: their conducting power, on this account, would be less; but on the other hand their elasticity *measured by the force required to compress them* is vastly greater, so that, as the table shows, water conducts sound better than air.

6. *But in the same medium velocity is uniform.*—The velocity of sound waves in air or in water, for example, is uniform. Moreover, all sounds in the same medium travel with the same velocity. When we listen to the music of a distant band, the various notes, high and low, loud and soft, reach the ear in the same order in which they were made. So also the shrill chirping of insects, the dull thud of a falling stone, the

melodious songs of the birds, and the murmur of rivulets, are all borne with equal swiftness through the air.

So uniform is the velocity of sound, that distances may be measured by means of it. Suppose the flash of a cannon on a distant hill was seen, and in 10 seconds afterward the report was heard, the temperature at the time being 61° F. The velocity of sound is 1,118 ft., and the sound waves, starting when the flash was seen, took ten seconds to reach the ear. $1,118 \times 10 = 11,180$. The observer was at distance of 11,180 ft. from the cannon.

§ 2. OF REFRACTION AND REFLECTION OF SOUND.

(54.) Sound waves will pass from one medium to another. In this case refraction of sound occurs. Sound may be made louder, by so refracting the waves that they will be collected at the place where the sound is heard.

1. *Sound waves pass from one medium to another.*—When in a room, with doors and windows closed, we are able to hear sounds distinctly that are made in the open air. The rattling of carriages, the singing of birds, and the voices of friends, come freely through the solid walls of our houses. To do this, the sound waves must pass from the air outside, into the solid material of the wall, and then from this again into the air of the room.

2. *Refraction of sound.*—Now, when sound waves go from one medium into another, they are *bent* out of the straight line in which they were moving; this is called *refraction* of sound.

To illustrate refraction: Suppose the lines B C and F H (Fig. 74) to represent two sound waves passing through air and striking the surface of another substance A A, at the points C and H. They will not go through in straight lines to E and L, but on entering the denser medium A A, they will be bent, taking the direction C D and H K, and when they emerge they will be again bent, so as to take the direction D E and K L. In this case, the refracted waves D E and K L are parallel to the original waves B C and F H. It will always be so when the sides of the medium, A A, are plane and parallel.

Fig. 74.

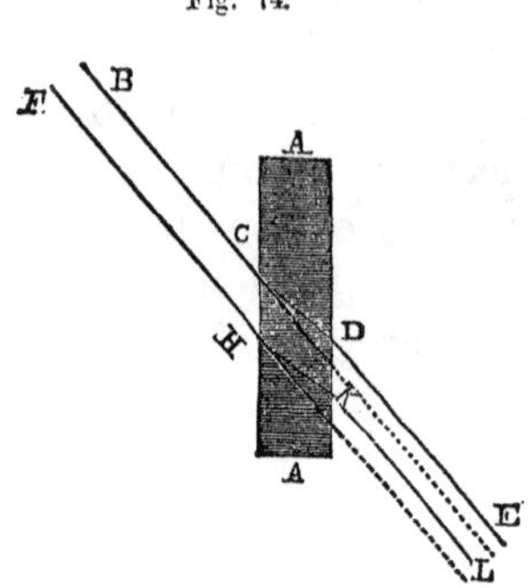

3. *Sound made louder by refraction.*—If the surfaces of the medium are curved instead of being plane and parallel, the sound waves which pass through will not come out parallel to those which enter. It may be that the waves which enter are *separating* from each other, and yet those that come out are *approaching* each other. In this way sound waves may be collected at a point, so that a sound may be heard there, which would not otherwise be audible.

This interesting fact may be illustrated by a curious experiment. A sack *a m n* (Fig. 75), made of two films of collodion, or of very thin india-rubber, united at their edges by a rim of iron, is filled with carbonic acid—a gas much denser than air. A watch is placed at W, near to the sack. If now a person put his ear at S, it may be, a point at a distance of five or six feet on the other side of the sack, the ticking of the watch will be heard distinctly. If the ear be moved

Fig. 75.

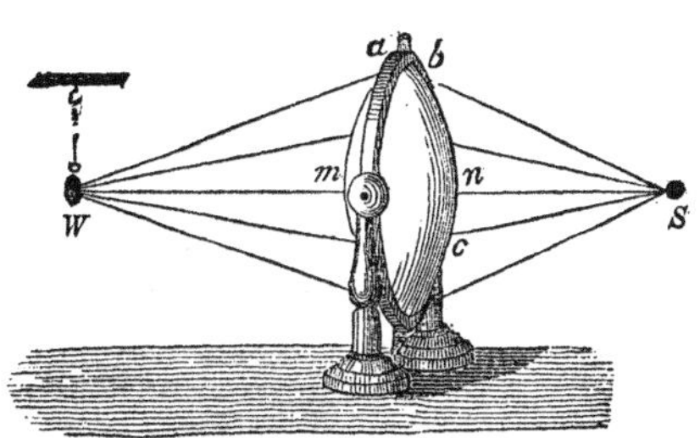

from this place ever so little, the sound will be more feeble, and if the sack be taken away the sound will not be heard at all.

This experiment beautifully illustrates refraction. The sound waves which start from the watch would go in straight lines outward, farther and farther apart. But they strike the surface of the sack at points *a m*, &c.; pass into it and through it, being bent from their course so as to emerge in the directions which take them all to the point S. So many vibrations are thus collected at this point that the sound is heard, whereas, if the sack were taken away, they would be so scattered that no sound would be produced.

(55.) When sound waves fall upon the surface of a second medium, only a part of them enter; the rest are reflected. The reflection of sound is governed by the following law:—

The angle of reflection must be equal to the angle of incidence.

An echo is produced by the reflection of sound.

1. *The reflection of sound.*—To illustrate the reflection of sound suppose the line I A (Fig. 76) to represent the direction of several sound waves, which, passing through the air, strike a body M M. Some of these waves will pass through the body, being refracted, but others will be thrown off in the direction A R. These are the reflected waves.

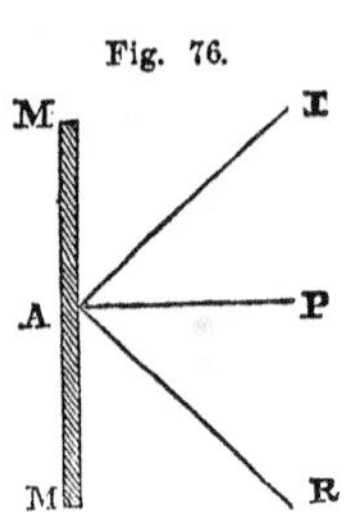

Fig. 76.

Now, a person standing at R will hear the voice of another at I, when the distance is considerable, sounding as though it came from a person in the direction R A. We always judge the direction of a sounding body from us, to be that *from which the waves enter the ear.*

2. *The law of reflection.*—To understand the language of this law, let us refer again to Fig. 76. The waves I A, those which fall upon the reflecting surface, are called the *incident* waves: the waves A R, those that are thrown off from the surface, are called the *reflected* waves, and the point, A, is called the *point of incidence.* Now, if a perpendicular A P, be drawn to the reflecting surface at the point of incidence, then the angle I A P, is the *angle of incidence*, and the angle P A R, is the angle of reflection. The *law* of reflection requires that these angles shall always be equal.

3. *The echo.*—An echo is a repetition of sound produced by the reflection of waves from a distant object. Who, after loudly uttering a word or sentence, has not

sometimes listened to the sound of his own voice coming back to him from a distant wood, or from the face of a cliff; or, it may be, from the wall of a distant building? Visitors to Cooperstown will not soon forget the fine echo returned from the rocky hills which skirt Otsego Lake. There, we are told by Fennimore Cooper, once dwelt Natta Bumpo, the hero in the story of the "Pioneers." Let his name be loudly called from a cer tain place upon the lake, and immediately the response—Nat-ta-Bum-po, every syllable full and clear, rings back over the water as if spoken by the hero himself from his cave in the cliffs.

When two obstacles are opposite to one another, the sound may be reflected back and forth many times. Surprising repetitions of echoes are, in this way, sometimes produced. It is said that an echo near Milan repeats a single sound thirty times. "When a trumpet is sounded at the proper place in the Gap of Dunloe, the sonorous waves reach the ear after one, two, three, or more reflections from the adjacent cliffs, and thus die away in the sweetest cadences."

§ 3. ON MUSICAL SOUNDS.

(56.) Musical sounds are caused by rapid vibrations which follow each other with great regularity. Any noise whatever, when repeated rapidly, will cause a continuous tone: even separate puffs of air, following each other rapidly, produce a musical sound.

1. *Musical sounds.*—When a single and intense air wave is suddenly produced, as when a gun is fired, the resulting sound is called a *report.* Let a series of such sounds be made in quick but irregular succession, and

the resulting sound is called *noise.* But when the waves are made with regularity, and follow each other so swiftly that the ear can distinguish no interval of time between them, the result is a *musical sound.*

2. *Any noise repeated rapidly causes a continuous tone.*—No matter what the source of the waves may be, nor how unmusical the separate noises, only let them be repeated with regularity and rapidity, and they will result in music. Slowly pass a piece of ivory, or even the finger nail, over the rough surface of a wound piano wire, and the sound of its strokes against the separate ridges is altogether unpleasant; but pass it *quickly* over the same surface, and the ear is saluted with a musical tone of surprising shrillness and purity. If a card be pressed against the teeth of a wheel which rotates slowly, a series of distinct and unpleasant taps will be heard; but, if by means of a larger wheel and band, this wheel be made to revolve rapidly, the taps will coalesce and salute the ear with music.

Fig. 77.

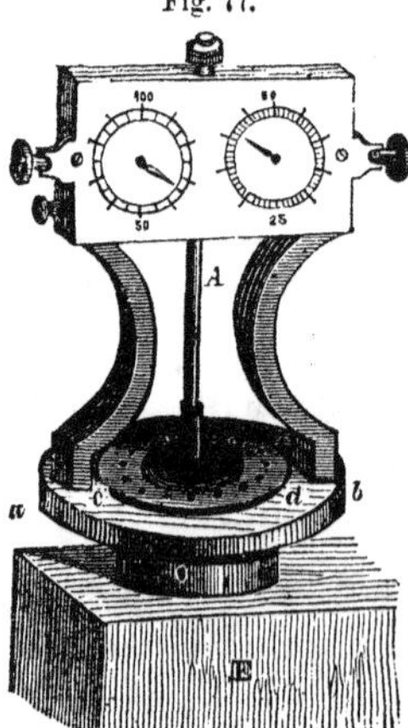

3. *Puffs of air made rapidly produce a musical sound.*—The *syren* is an instrument by which a series of air puffs are made to produce a musical sound, and by which the number of puffs made in a second are registered. Its structure may be learned from Fig. 77.

A brass tube O, leads from a wind chest E, to a brass plate, *a b*, which is pierced with a series of holes arranged around the circumference of a circle.

Above this plate is a disk *c d*, also perforated with holes exactly corresponding to those in the plate below. The disk is provided with a steel axis A, and is so fixed that it may rotate with a very small amount of friction. The wheel work shown in the upper part of the figure registers the number of puffs made in any given time.

Now, when the disk *c d* revolves, the holes in it will be brought alternately over the perforations in the plate *a b*, and the spaces between them, so that these holes will be alternately opened and closed. When the disk is still, and the holes are open, if air be urged through the tube O, it will escape from the top in steady streams, but when the disk revolves these streams will be cut up into successive puffs. If the disk turns slowly, the separate puffs are heard, but as the disk is turned more and more rapidly, the air announces its escape by a musical sound of great purity and increasing shrillness.

By a simple artifice, the air which gives the sound is made to turn the disk. This is done by making the holes through the plate *a b*, oblique instead of vertical; those in the disk being also oblique, but inclined in the opposite direction.

(57.) Musical sounds differ in three respects: 1st, Pitch; 2d, Intensity; and 3d, Quality.

I.—PITCH.

A.—Pitch depends entirely upon the rapidity of vibrations which produce the sound.

The difference in the pitch of two sounds, is called an interval, and a series of eight sounds of different pitch, has been adopted as the foundation of all music, and called the diatonic scale. The number of vibrations to produce the note *a* of the treble clef, is 440 a second.

1. *Pitch depends on the rapidity of vibration.*—The pitch of sounds is that which distinguishes them as being high or low. It depends entirely upon the rapidity of vibration: the more rapid the vibrations, the higher will be the sound produced. Two sounds made by the same number of vibrations per second, however much they may differ in other respects, will have the same pitch.

2. *Intervals.*—When the number of vibrations which produce one sound, is twice as great as that which produces another, we must not say that the sound is *twice as high*, but rather that it is an *octave above*. The term octave, is used to designate a tone which is made by twice the number of vibrations needed to produce a lower one, called the *fundamental*. Other intervals will be named in the description of the scale.

3. *The diatonic scale.*—Now the difference in pitch, or the interval between a fundamental note and its octave, is very great. To fill up this interval, sounds have been chosen which blend, or harmonize most perfectly with the fundamental, or with each other. These, placed between the fundamental and its octave, form a series of eight notes, called the *natural*, or the *diatonic* scale.

The eight notes of the scale are expressed by the following names and intervals:—

Names,	C,	D,	E,	F,	G,	A,	B,	C.
Intervals,	1st,	2d,	3d,	4th,	5th,	6th,	7th,	8th.

This scale repeated about eleven times, making what is termed in music eleven octaves, will include all sounds within the range of the human ear. Only about seven octaves are available in music.

The method of representing the notes in music is familiar to all. Remember that the note called A, *is found in the second space of the treble clef*, and the position of all others may be easily traced.

4. *Number of vibrations for the notes.*—The number of vibrations to produce the various notes, may be found by experiment with the syren. (See Tyndall on Sound.) It has been found that the note A, of the treble clef is made by 440 *vibrations a second.* (See Silliman's Phys.) In piano-fortes for private use, this note is produced by about 420 vibrations a second.

If we represent the number of vibrations for the fundamental note by 1, then the several notes of the scale will be made by the following ratios :—

C,	D,	E,	F,	G,	A,	B,	C.
1,	$\frac{9}{8}$,	$\frac{5}{4}$,	$\frac{4}{3}$,	$\frac{3}{2}$,	$\frac{5}{3}$,	$\frac{15}{8}$,	2.

Now, remembering this series of fractions, and the fact that A is made by 440, the number of vibrations for all the others may be found. Thus for example, how many vibrations to give the fundamental C? The *relative* number of vibrations for A and C are $\frac{5}{3}$ and 1; that is, A is produced by $\frac{5}{3}$ as many vibrations as C; or, to reverse the ratio, C requires $\frac{3}{5}$ as many as A, and $\frac{3}{5} \times 440$ 264. Having this number for the fundamental, multiply this by the fractions $\frac{9}{8}$, $\frac{5}{4}$, $\frac{4}{3}$, &c., and the numbers for the corresponding notes will be obtained. These multiplied by 2 will give the number for the notes in the next higher octave, or divided by 2, will give the numbers for the notes in the octave below.

II.—INTENSITY.

B.—The intensity of sound is that which distinguishes it as being loud or soft. It depends *entirely* upon the *amplitude* of the vibrations which produce it. The greater the amplitude, the louder the sound will be. In the case of a vibrating string for example, the *loudness* or intensity of the sound made by it, will depend entirely upon the distance through which the string vibrates across its line of rest.

III.—QUALITY.

C.—By quality, we refer to that peculiarity of sound by which we may distinguish notes of the same pitch and intensity, made on different instruments. The pitch and intensity of notes made on a violin and on a piano may not differ, and yet how easy to tell the sounds apart. We recognize the voices of friends, not by their pitch nor their intensity, but by their *quality*.

Quality is thought to depend upon the different sets of vibration, which, in different instruments, combine with those that cause the leading tone. The material of a violin vibrates as well as the string which is stretched upon it, and the sound made by both sets of vibrations is the real tone of the instrument. The various parts of a piano vibrate as well as the piano wire, and the sound produced by all these vibrations together is the familiar sound of the instrument. Now, it is clear that the sets of vibration in these two instruments must be different, and for this cause the *quality* of the two tones is different.

(58.) Musical instruments are, for the most part, of two classes: first, those in which the sounds are produced by vibrating strings, and, second, those in which sounds are made by vibrating columns of air.

I.—STRINGED INSTRUMENTS.

A.—In stringed instruments, the pitch of the different notes is obtained by using strings or wires of different lengths, of different tensions, and of different weights.

1. *Stringed instruments.*—The violin, the guitar, and the piano, are familiar forms of stringed instruments. In every case, cords or wires are tightly stretched over some solid body having considerable surface. The music of these instruments is not made by the vibrations of their cords alone; the simple vibration of a cord is not able to produce sound of sufficient intensity, but by being stretched over hollow boxes made of elastic wood, the material of the box, and the air inside of it are made to vibrate, and these vibrations, joined with those of the cords, produce the sounds of the instrument.

2. *Pitch varied by using strings of different lengths.*—The pitch of any sound depends upon the rapidity of vibrations; but according to the first law of vibrating strings, the rapidity of vibration is greater as the string is made shorter. To obtain sounds of different pitch, we may then use strings of different lengths.

Now, suppose we would know the lengths of eight strings of the same weight and tension, which would give the eight notes of the scale. We have learned that the number of vibrations per second is *inversely* as the length of the cord, and we have learned also that the relative number of vibrations for the eight notes are expressed by the series 1, $\frac{9}{8}$, $\frac{5}{4}$, $\frac{4}{3}$, $\frac{3}{2}$, $\frac{5}{3}$, $\frac{15}{8}$, 2. Then *invert* the terms of this series, and they must express the relative lengths of cord to produce the notes. They will be 1, $\frac{8}{9}$, $\frac{4}{5}$, $\frac{3}{4}$, $\frac{2}{3}$, $\frac{3}{5}$, $\frac{8}{15}$, $\frac{1}{2}$. Knowing the length of the string to give the fundamental, it is easy to calculate the lengths of all the others. Let us start with a string 18 inches long for the first note; the second must be $\frac{8}{9} \times 18$; the third must be $\frac{4}{5} \times 18$; the fourth, must be $\frac{3}{4} \times 18$, and so on until the eighth, which must be $\frac{1}{2} \times 18$.

3. *Pitch is varied by using strings of different tension.*—According to the second law of vibrating strings [see (46.)], the number of vibrations made in a second increases when the tension increases. Hence the pitch of sound made by the string will be higher when the tension is made greater.

4. *Pitch is varied by using strings of different weights.*—According to the third law of vibrating strings, the number made in one second varies inversely as the square root of the weight of the strings. Hence the pitch of the sound will be higher, when the string which makes it is lighter.

II.—WIND INSTRUMENTS.

B.—The organ and the clarionet are examples of wind instruments. In the organ, sounds are made by vibrating columns of air in pipes, sometimes aided by the vibrations of a slender and elastic tongue, called a reed. (See Tyndall on Sound.) In the clarionet the sounds are always made by air vibrations aided by a reed.

The pitch of sounds in pipes, depends upon the lengths of the pipes. A pipe to produce the *lowest* note in music, must be 32 ft. in length, and the pitch of tones from other pipes will vary inversely as the length of the pipes.

Organ pipes are sometimes open at the top, and sometimes closed. An open organ pipe yields a note an octave higher than a closed pipe of the same length. A closed pipe, to give the lowest note in music, need only be 16 ft. in length.

CHAPTER VI.

THE EFFECTS OF VIBRATIONS.—II. LIGHT.

§ 1. ON THE NATURE OF LIGHT, AND THE LAWS OF ITS TRANSMISSION.

(59.) READ (51). Light is thought to be the effect of vibrations in an ether which fills all space not filled by other matter. These vibrations are produced by luminous bodies, and, when transmitted to the eye, cause vision.

1. *Light is the effect of vibrations.*—It was once thought that light consists of minute particles of matter thrown in great abundance from the sun and some other bodies: it is now generally believed that light is the result of vibrations. But light will pass through the most perfect vacuum that can be made: what can be left to vibrate? Moreover, the atmosphere extends but a few miles above the earth, yet the light from the sun comes in floods through the vast distance which separates these bodies: what can there be between the sun and the earth, whose vibrations bring to us the sunlight?

2. *The ether.*—Philosophers assume that there is a thin, elastic substance called ether, much finer and rarer than air, which fills all the spaces between the heavenly

bodies, and enters into all the spaces between molecules of matter in every form. The vibrations of this ether carry light wherever it goes, through a vacuum, through celestial spaces, through bodies like glass, and through the substances of the eye, until it strikes the nerves of sight.

When a gas jet is suddenly lighted in a dark room, every eye present is dazzled by the brightness of the light. The explanation is this. The heated gas makes the ether vibrate. This ether is between the particles of the air, and between the particles of the eye. Its vibrations, starting from the gas jet, go through the air and into the eye, and when they reach the delicate nerves in the back part of this organ we are made conscious of the presence of light.

3. *Luminous bodies.*—Bodies which shine by their own light are called *luminous* bodies. They are bodies which can make the ether vibrate. Bodies which shine only by light which they receive from others are called *non-luminous* bodies. They can not make the ether vibrate. The sun is a luminous body: so is a red-hot iron ball. All flames are luminous bodies. The moon is non-luminous. Almost all bodies on the earth are non-luminous: the light which they give to us is light which the sun first gave to them.

(60.) Rays of light are transmitted through some media more freely than through others but always according to two laws:—

1st. In a medium of uniform density, light goes in straight lines with a uniform velocity.

2d. The intensity of light varies inversely as the square of the distance from its source.

The art of Photometry depends upon this second law.

1. *Rays of light.*—A single line of light, or more accurately the path of a single vibration, is called a *ray* of light. But the smallest portion of light which can be separated by experiment, consists of many rays. A ray of light is quite too delicate a thing to be seen. A collection of parallel rays is called a *beam* of light. A collection of rays which diverge from a point, or which converge toward a point, is called a *pencil* of light.

2. *Rays of light are transmitted.*—Some substances permit light to pass through them freely; they are said to be *transparent.* Air and water are examples of transparent bodies. Others, such as iron and wood, appear to forbid the passage of light through them: they are said to be *opaque.* But no substance will transmit all the light which it receives; even the air is not perfectly transparent. On the other hand, no substance will stop all the light which falls upon it; even gold, when a very thin leaf of it is examined, can be seen to transmit light. All substances are doubtless able to transmit light in some degree.

3. *Light moves in straight lines.*—That light moves in straight lines is shown by numerous familiar facts. We can not see through a crooked tube, simply because light can not pursue a crooked path. And again: who has not seen the sunlight coming through the shutters of a half-darkened parlor, spotting the opposite wall with circles of light? The sun, the hole in the shutter, and the spot on the wall, are always in the same straight line. Let the air of the room be sprinkled with dust,

and the paths of the sunbeams are seen streaking the air with bars of light.

4. *With uniform velocity.* — Light travels through space with a uniform velocity of about 192,000 miles a second. This number has been found by observing the eclipses of one of the moons of Jupiter. The time when the eclipse should begin can be calculated by an astronomer with great accuracy. But it is found that when the earth is in that part of its orbit nearest to Jupiter, the eclipse begins 16 minutes and 36 seconds sooner than it appears to when the earth is in the opposite part of its orbit. It must, therefore, take light 16 minutes and 36 seconds to go across the earth's orbit. When this distance is known and divided by the number of seconds the velocity of light is found. Calling the distance from the earth to the sun, ninety-five millions of miles, the result is about 192,000 miles a second.* For all distances on the surface of the earth, the passage of light may be considered instantaneous. It would go quite around the world almost seven times in a single second.

5. *The second law.*—That the intensity of light is less as we go farther from the luminous body, is a fact familiar to all: the *rate* at which it diminishes is not so apparent. That the intensity varies inversely as the square of the distance may be easily proved by experiment.

A square piece of stiff card-board A (Fig. 78), is placed in front of another (B) very much larger. If now, a candle-flame be placed in front of the small card, a shadow will be cast upon the large one. This shadow

* For Foucault's method of finding the velocity of light, see Silliman's Phys. p. 294.

will be larger, as the small card is moved nearer to the flame, it will be smaller as it is moved the other way.

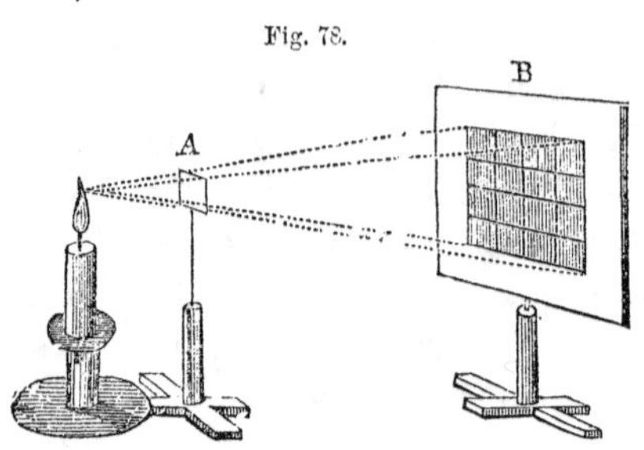

Fig. 78.

The figure is intended to show the small card to be just $\frac{1}{4}$ as far from the flame as the large one. In this case the shadow will be found to be exactly 16 times as large as the card in front of it. Now, the same amount of light which is spread over the small card would, if it could go on, just cover the place of this shadow. But if the same amount of light is spread over 16 times as much surface in one case as in another, it can be only $\frac{1}{16}$ as intense. At 4 times the distance from the luminous body, in this case, the intensity of the light is $\frac{1}{16}$ as great. At 3 times the distance, the light would, in the same way, be found to be $\frac{1}{9}$ as intense. In other words; the intensity of light varies inversely as the square of the distance from the luminous body.

6. *Photometry.*—It is often desirable to compare the *illuminating powers* of different flames. The art of doing this is called *photometry.* The simplest method is to place the two flames at such distances from a screen, that the intensities of the light they shed upon it shall be equal; the illuminating powers of the flames must then be as the squares of these distances. Suppose, for example, that we wish to know how many times more light one candle will give than another of inferior quality. Let a slender rod B (Fig. 79), be put just in

front of a white screen A, and then move the flames to such distances, that the two shadows of the rod, falling

Fig. 79.

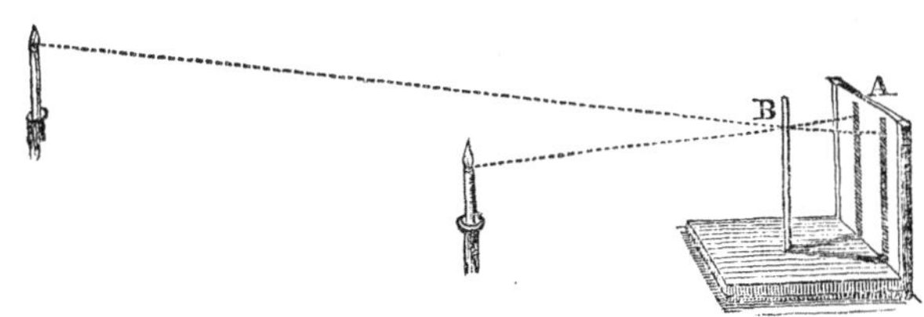

side by side upon the screen, shall appear to be of *equal darkness.* The intensities of their lights on the screen must then be equal. Measure the distances from the flames to the screen: the amounts of light they give will be as the squares of their distances. One being twice as far away as the other, it gives four times as much light.

§ 2. ON THE REFLECTION OF LIGHT.

(61.) When light in passing through one medium comes against the surface of another, only a part will be transmitted, another part will be reflected, obeying the following law:—

The angles of incidence and reflection must be equal, and in the same plane.

1. *Reflection.*—The reflection of light is in all respects like the reflection of sound [see (55.) 1]. The same terms are used to describe it; the same figure may be reproduced to illustrate it. Thus in Fig. 80, the line I A, may represent a beam of light passing through air and striking upon the surface of a plate of glass at A. One part of the

Fig. 80.

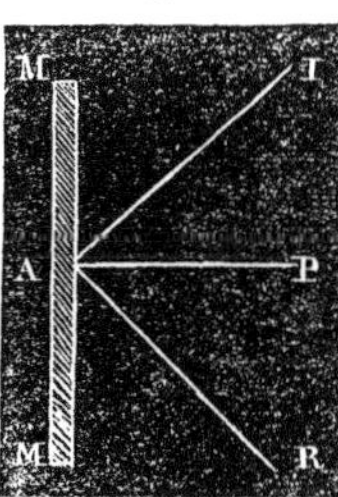

beam will enter the glass and emerge again on the other side, but another part will be thrown back into the air in the direction A R. The beam I A is the incident beam. The beam A R is the reflected beam. The point A is the point of incidence.

2. *The law of reflection.*—The reflection of light is also governed by the same law as the reflection of sound. The angle I A P (Fig. 80), is the angle of incidence. The angle P A R is the angle of reflection. These two angles must be equal.

How various and beautiful are the phenomena which this principle of reflection explains! The sky, with all its floating clouds or shining stars, is painted in every pool of water, because the light from them, falling on the surface of the water, is reflected to our eyes. Rocks, and shrubbery, and dwellings along the shore, are pictured in the quiet waters of the lake, with skill exceeding that of any human artist.

Vision is produced by reflected light. How seldom do we receive the direct rays of the sun into the eye; how rarely indeed, do we look directly upon any luminous body! But in all other cases we see objects only by reflected light. The sunbeams fall upon all objects exposed to them, and, bounding from their surfaces, enter the eye, and we see them in the direction from which the reflected rays have come.

(62.) The effects of mirrors are explained by reference to the law of reflection.

Rays of light reflected by a plane mirror have the same relation to each other as before reflection;

But the effect of a concave mirror is to collect the rays of light which are reflected by it;

While a convex mirror always separates the rays which it reflects.

1. *Mirrors.*—Any surface smoothly polished that will reflect nearly all the light which falls upon it, is called a *mirror.* The smooth surface of quiet water is a very perfect mirror. Artificial mirrors are generally made of metal or of glass. If made of glass, a thin film of mercury is spread over one side, and the smooth surface of this metallic coating is really the reflecting surface. Mirrors are either *plane* or *curved.* Of the curved mirrors there are two varieties, the *concave* and the *convex* mirrors.

2. *The effect of plane mirrors.*—The rays of light which fall upon a mirror may be parallel, or converging, or diverging, but can have no other relation. Now, let the mirror be represented by the straight line A B (Fig. 81), and suppose, first, that it receive two parallel rays represented by the lines *a c* and *b d*. At the point of incidence *c*, erect a perpendicular to the surface A B. The angle, *a c g*, will be the angle of incidence. Then draw the line *c f*, so as to make the angle of reflection, *g c f*, equal to the angle of incidence, and *c f* must be the direction of the ray reflected from the point *c*. Again: at the point of incidence *d*, erect a perpendicular and draw the line *d e*, making the angle of reflection equal to the angle of incidence, and this line must represent the ray reflected from the point *d*. It will be found that the reflected rays, *c f* and *d e*, will be parallel.

Fig. 81.

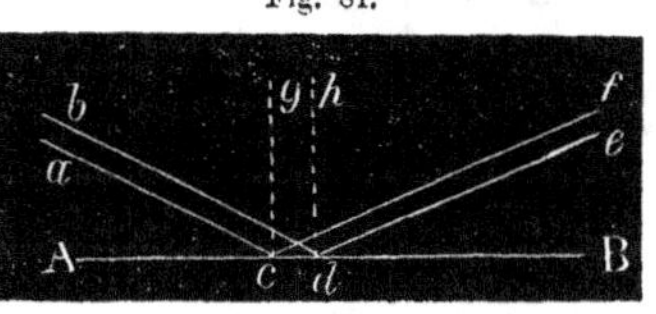

Fig. 82.

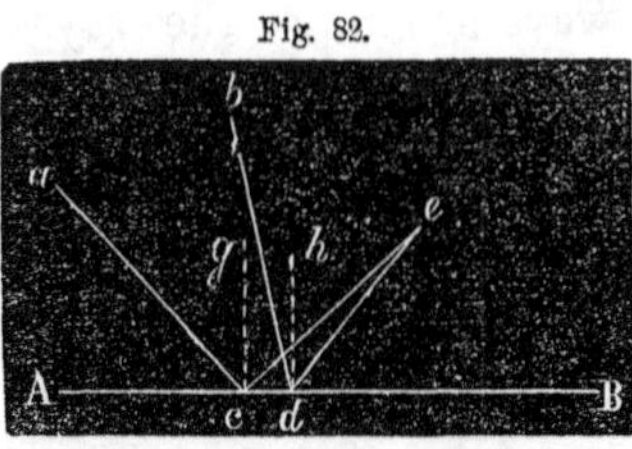

Suppose, second: that two rays, *a c* and *b d* (Fig. 82), are *converging* and strike the mirror at the points *c* and *d*. By making the angles of incidence and reflection equal, exactly as it was done in the preceding case, we find that the reflected rays will take the directions *c e* and *d e*, *converging* to the point *e*.

Fig. 83.

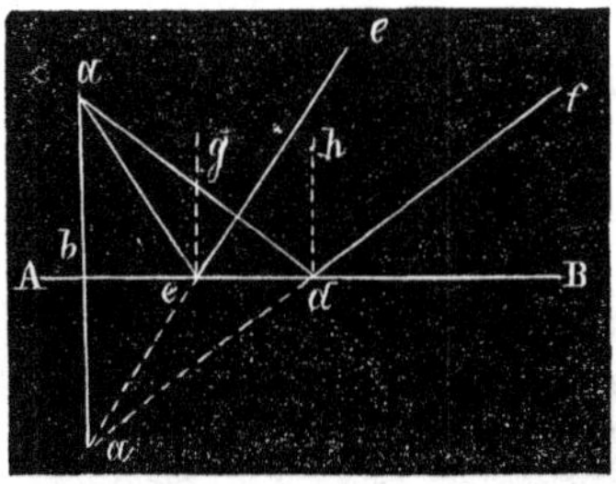

Suppose, third: that the rays are *diverging*. Represent them by lines *a c* and *a d* (Fig. 83). Erect the perpendiculars and construct the angles of incidence and reflection equal, and the directions of the reflected rays will be *c e* and *d f*, *diverging* from each other.

In each of these three cases, the reflected rays have the same relation as the incident rays.

3. *The effect of concave mirrors.*—We will notice only those concave mirrors whose surfaces are spherical. If we know the direction of the incident rays, we can find the direction of the reflected rays by making the angle of reflection equal to the angle of incidence. To construct the angle of incidence, we must, as in the plane mirror, erect a perpendicular to the concave surface at the point of incidence, and all difficulty disappears when we remember that *a perpendicular to any spherical surface is the radius of the sphere.*

Fig. 84.

In Fig. 84, M N represents a section of a concave mirror. The point C represents the *center of curvature*, that is, the center of the hollow sphere, of whose concave surface the mirror is a part. Now, if E A and D B represent two *parallel* incident rays, and we would find the direction they take after reflection, we may draw the radii, C A and C B, making the angles of incidence, E A C and D B C, and then draw the lines A F and B F, so as to make the angles of reflection equal to these. By so doing, we find that the reflected rays *converge* and cross each other at the point F.

Fig. 85.

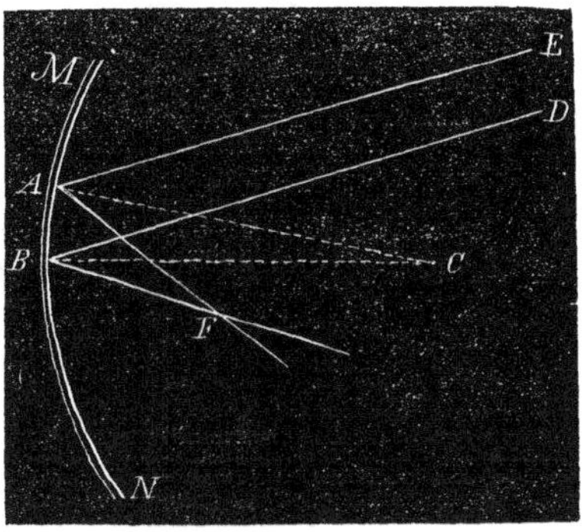

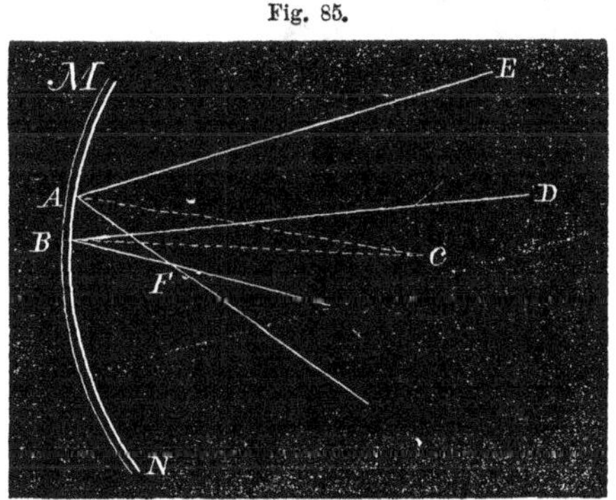

In Fig. 85, the lines, E A and D B, represent *converging* rays. By constructing the angles of incidence and reflection in the same way as before, we find that the reflected rays cross each other at the point F, *converging faster* after reflection than before.

In the same figure, F A and F B may represent *diverging rays*, striking the mirror at the points A and

B. By constructing the angles of incidence and reflection equal, we find the reflected rays taking the directions A E and B D, *diverging less* after reflection than before.

Now, since parallel rays are made converging, and converging rays are made more converging, while diverging rays are made to diverge less, we may say that the general effect of a concave mirror is to *collect* rays of light.

A *focus* is any point where rays of light cross, or appear to cross, after reflection. The points F, in Figs. 84 and 85, are foci. The *axis* of a mirror is a straight line drawn through the center of curvature and the middle point of the mirror.

Fig. 86.

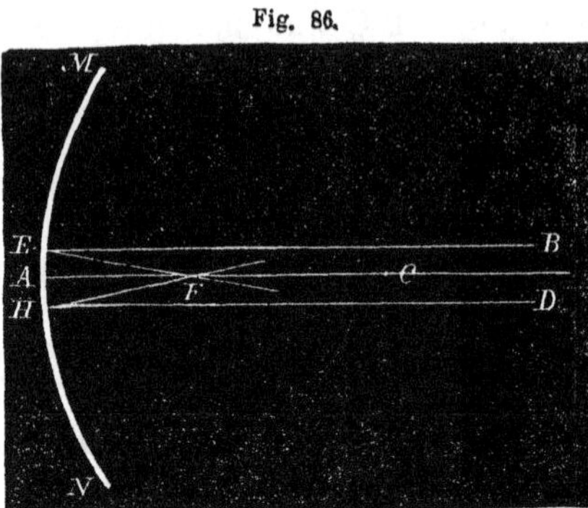

In Fig. 86, the line C A is the axis of the mirror M N, whose center of curvature is at C. The focus of rays that are parallel to the axis, and fall upon the mirror near its middle point, is called the *principal focus.* If the rays B E and D H (Fig. 86), are near to, and parallel to the axis C A, they will, after reflection, cross each other at the point F, and this point is the *principal focus* of the mirror. The principal focus is on the axis, half way between the center of curvature and the mirror.

4. *The effect of convex mirrors.*—In Fig. 87, a convex

mirror is represented by M N, its center of curvature by the point C. Two parallel rays of light, E A and D B, strike the mirror at the points A and B. To

Fig. 87.

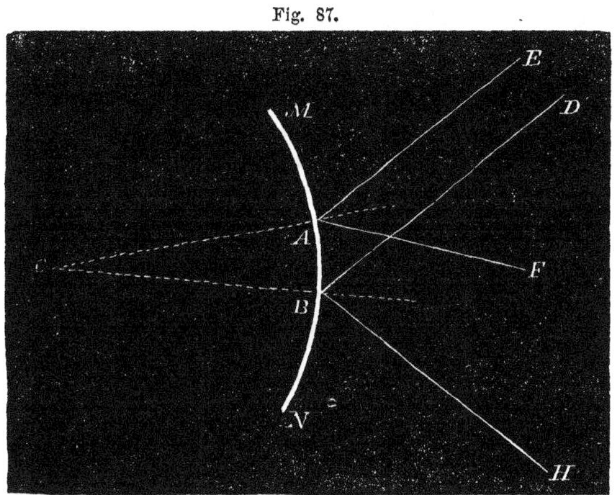

construct the angles of incidence, we must erect perpendiculars to the surface at these points. The perpendiculars are the radii, C A and C B, *extended* beyond the convex surface of the mirror. By making the angles of reflection equal to the angles of incidence, the reflected rays are found to take the directions A F and B H. We notice that parallel rays are rendered diverging.

So we might show that diverging rays would be made more diverging, and that converging rays would be made to converge less. We say, therefore, that the general effect of a convex mirror is to *separate* rays of light.

(63.) When the light reflected from a mirror enters the eye, we see an image of the object from which the light proceeds.

The image of any point will always be found where the rays of light which go from that point, either meet, or appear to meet, after reflection.

1. *Images by reflection.*—When a person stands before a looking-glass, numberless rays of light from every point of his countenance fall upon it. These rays are reflected, and many of them are thrown into the eye. Those which enter the eye, cause him to see his image in the glass.

2. *The image of a point.*—Now, if the rays of light, which form the image in the glass, were visible, the person would be able to trace them back from the eye, converging toward the points on the glass from which they are reflected, and they would appear as if they came from points in the image behind the glass.

Fig. 88.

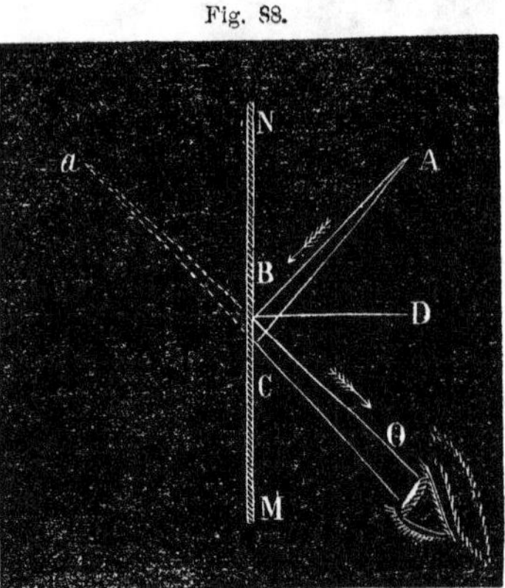

This will be understood by means of Fig. 88. Let M N represent a plane mirror. From the point A, numberless rays fall upon the mirror, some of which, after reflection, will enter the eye, supposed to be at O. Two of these rays are represented in the figure. The eye will receive these rays as if they came from the point *a*, and this point *a* is the image of the point A, from which the rays proceed.

(64.) The image formed by a plane mirror is always

as far behind the mirror as the object is in front of it; the same size as the object, and erect.

1. *Images by plane mirrors.*—We are now prepared to see how looking-glasses make such perfect images of all objects placed in front of them. Suppose an arrow A B, placed before a mirror (Fig. 89). Let us construct its image. From the vast number of rays which go from A to the glass, select two which fall upon it very near together, at *f* and *g*. By making the angles of reflection equal to the angles of incidence, we find the reflected rays taking the directions *f* P, and *g* O. Now, if the eye be placed at E, it will receive these reflected rays as if they came from the point *a*. Again, select two rays, which, going from the other end of the arrow B, strike the mirror at points near together at *c* and *d*, so that after reflection they can enter the same eye at E. These rays will appear to have come from *b*. From all points between A and B, rays of light will go to the mirror; and, being reflected, will enter the eye at E, and appear to have come from points between *a* and *b*. The image of the arrow, A B, will thus be seen at *a b*. We may describe this image thus: *the image made by a plane mirror is always behind the mirror, just as far as the object is in front of it, of the same size as the object, and erect.*

Fig. 89.

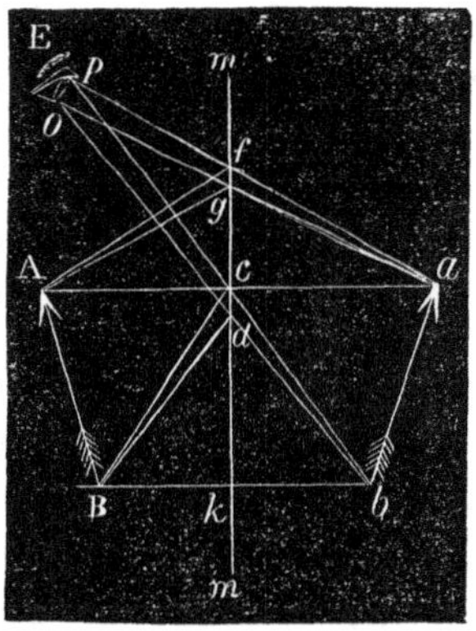

(65.) If an object be placed in front of a concave

mirror, an inverted image may be formed on the same side of the mirror. To explain this, remember that the image of any point will be, where rays of light either meet, or appear to meet, after reflection.

1. *Images are formed.*—The brilliant inner surface of a silver spoon shows the image of a person who looks upon it, but it will be curiously different from his image seen in a looking-glass. It is very small; it is inverted; and, moreover, by careful attention, the person sees his picture standing in the air between himself and the surface of the spoon. Nor is this all; the picture in the air will grow larger or smaller, or it may disappear altogether, as the spoon is moved toward or from the face of the observer.

If a spherical concave mirror of small curvature be at hand, a beautiful experiment will illustrate its power to form images. Let a beam of sunlight pass through an opening in the shutter of a dark room. In the path of this beam, at a convenient distance from the window, place a picture of a butterfly or other object, painted in transparent colors upon glass. The concave mirror placed in front of the picture, so as to receive the light which has come through it, will reflect the rays upon the wall above the window, and if its distance from the picture is just right, a magnificent image of the butterfly, much larger than the picture, and with its head downward, will be seen upon the wall.

2. *The images of points.*—How is this beautiful effect produced? Can we find the images of points [see (63.) 2] of the object by tracing the reflected rays which produce them? Let M N (Fig. 90), represent a section of a concave mirror, and suppose an arrow, A B, in front of it. Select two rays of light which, going

from the point A, fall upon the mirror at the points D and E. After reflection they will cross each other

Fig. 90.

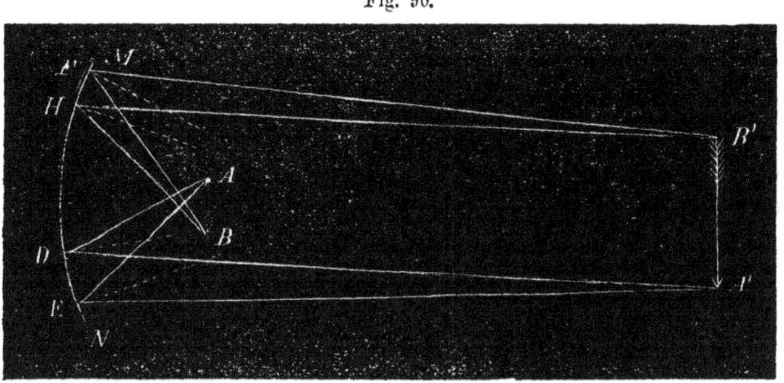

at A′. Again select two rays, which, going from the point B, fall upon the mirror at the points H and F. After reflection they will cross each other at B′. Other points in the object will send rays to the mirror, which, after reflection, will cross each other at points between A′ and B′. In this way a large and inverted image is made in the air at A′ B′.

(66.) The position and size of the image will depend upon the distance of the object from the mirror. We will notice three well-marked cases :—

1st. When the object is beyond the center of curvature.

2d. When the object is between the center of curvature and the principal focus.

3d. When the object is between the principal focus and the mirror.

1. *The object beyond the center.*—We are now prepared to see how the mirror forms its images in the air. Let M N (Fig. 91) represent a section of a concave

mirror, whose center of curvature is C, and whose principal focus is F. Suppose an arrow A B, to be

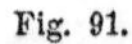

Fig. 91.

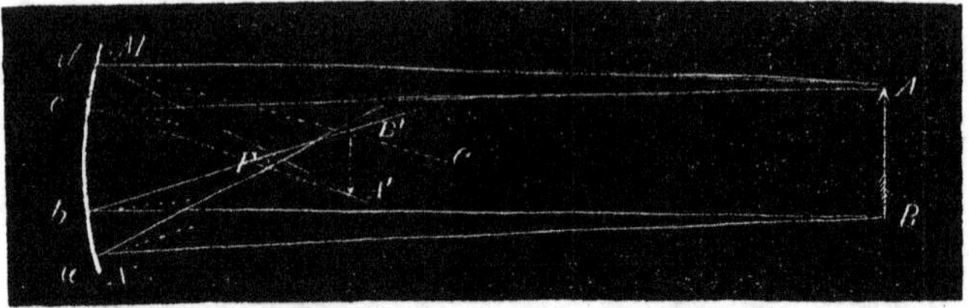

put in front of the mirror, beyond the center of curvature. The rays of light from the top of the arrow A, will, after reflection from the mirror, cross each other at the point A′. Those which go from the bottom of the arrow B, will, after reflection, cross each other at B′. From points of the arrow between A and B, the light which falls upon the mirror, will be collected into corresponding points between A′ and B′. A perfect image of the arrow will thus be formed at A′ B′. In this case we observe that *the image is between the center of curvature and the principal focus, inverted, and smaller than the object.*

This case was illustrated by the experiment with the silver spoon.

2. *The object between the center and focus.*—Now, let us suppose that in this same Fig. 91 an arrow B′ A′, with its head pointing downward, is placed between the center of curvature and the principal focus. The rays of light from the point B′, striking the mirror at *a* and *b* will, after reflection, cross each other at the point B, those from A′, after reflection, will cross each other at A, and the image of the arrow will be formed at A B. In this case, we observe that *the image will*

be beyond the center of curvature, inverted, and enlarged.

This case was illustrated by the experiment with the picture of the butterfly.

3. *The object between the focus and the mirror.*—When the object is gradually moved from the center toward the focus, the image will rapidly move farther and farther away, until, when the object has reached the focus, the image will be at an infinite distance *in front of the mirror*, and of course, invisible. But let the object be carried a little farther, so as to be between the focus and the mirror, and the image suddenly leaps from its distant place in front of the mirror, to a position *behind* it.

To illustrate the formation of this image behind the mirror, let A B (Fig. 92), represent an object between the focus F, and the mirror M M′. Two rays of light from the top of the object strike the mirror at H, and are reflected to the point E. To an eye placed there, these rays would appear to have come from the point *a* behind the mirror. Two rays from the bottom of the object falling upon the mirror at K, will be reflected so as to enter the same eye at E, and will seem to have come from the point *b*. Joining the points *a* and *b*, we have the entire image constructed. In this case we observe that *the image is behind the mirror, erect, and larger than the object.*

Fig. 92.

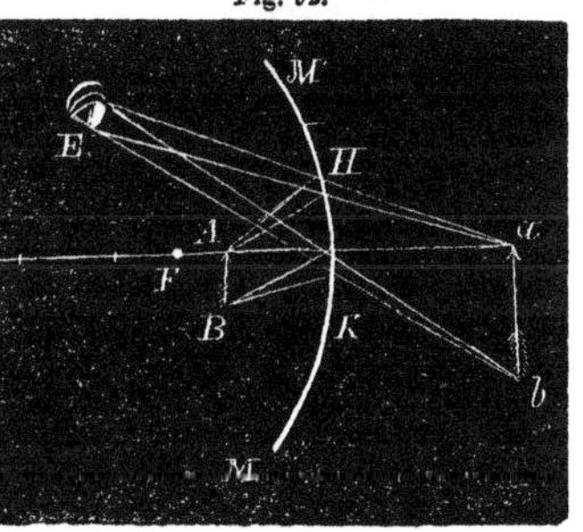

(67.) The images formed by convex mirrors are

always behind the mirror, erect, and smaller than the object.

1. *Images by convex mirrors.*—The bottom of a silver spoon will serve, in a homely way, to illustrate the effects of a convex mirror. A person looking upon it will see his own image, apparently in the metal of the spoon, erect, but very small. The following diagram will illustrate the formation of these images.

Fig. 93.

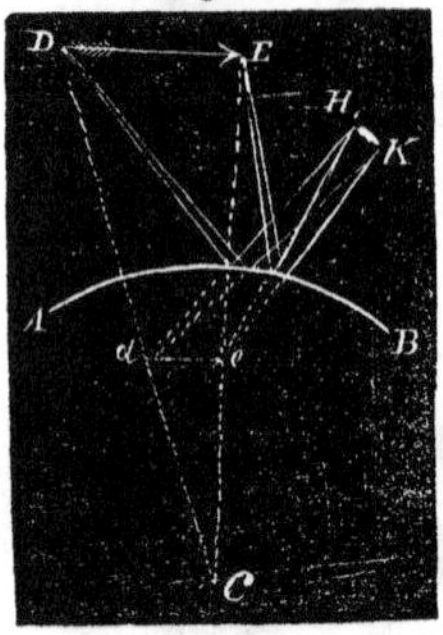

The object D E, is placed in front of the convex mirror A B, whose center of curvature is at C. Two rays of light from D, may be traced after reflection to the points H and K, and if an eye can receive these rays, they will seem to come from the point *d.* In like manner, rays from the point E, after reflection from the mirror, may enter the same eye, and appear to have come from *e.* The image of the object will thus be found at *d e*, *behind the mirror*, *erect*, and *smaller than the object.*

§ 3. ON THE REFRACTION OF LIGHT.

(68.) When light passes from one medium into another of different density, it is refracted, obeying the following laws:—

1st. In passing into a denser medium, light is bent toward a perpendicular to the surface at the point of incidence.

2d. In passing into a rarer medium, light is bent from the perpendicular.

1. *Refraction.*—The refraction of light is similar to

the refraction of sound [see (54.) 2]. The same terms are used to describe it, and the same figure might be made to illustrate it. A simple experiment will suit our purpose better. Through a small opening in the shutter of a darkened room, let a beam of sunlight enter, and fall obliquely upon the surface of water held in a glass vessel (*a b c d*, Fig. 94). If the water has been made turbid by the addition of a little soap, and the air above it misty, by sprinkling into it the dust of a chalk-brush, the beam of light will be distinctly seen in both, absolutely straight, except at the surface of the water, where it will be very considerably bent. Its path is represented by the broken line A B D.

Fig. 94.

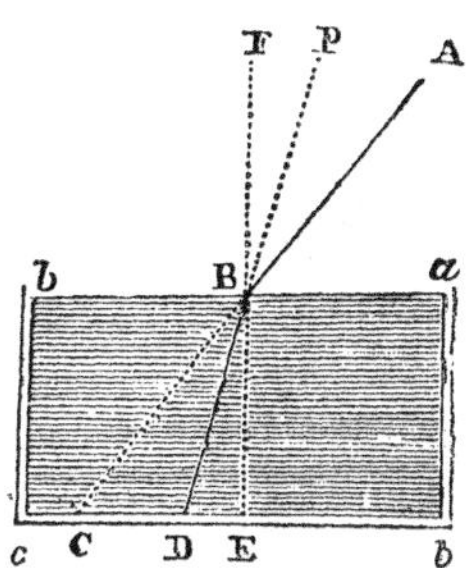

2. *The first law of refraction.*—If now, a perpendicular F E, be erected to the refracting surface at the point of incidence B, we see that the rays A B, instead of moving in a straight line, onward to C, will be bent toward the perpendicular. Water is denser than air. In going from the rarer to the denser medium, the light is bent *toward* the perpendicular.

3. *The second law of refraction.*—Let us suppose that D B represents a beam of light going from the water into the air at B; it will take the direction B A, instead of going on in a straight line toward P, being bent from the perpendicular F E. In passing from the denser medium into the rarer, the light is bent *from* the perpendicular.

Many phenomena in nature may be explained by

reference to these principles. When, for example, an oar is dipped into clear and quiet water, it appears broken at the surface. The light comes to the eye from all points of the oar. From that part which is above water it comes in straight lines through the air, but from the part under the water the light coming up into the air, is bent at the surface. The eye which receives these bent rays traces them back in *straight lines*, and the oar, from which they come, is thus made to appear to be where it really is not.

(69.) Some substances refract light more than others. Their relative refracting powers are indicated by certain numbers, which are called indices of refraction.

1. *The index of refraction.*—We may best explain the meaning of this term by means of the following diagram. Suppose a small beam of light L A (Fig. 95), to be passing from air into water. It will be bent at A, and go on in the direction of A K. Now, with the point A as a center, and with any convenient radius describe a circumference. Let a perpendicular B C, be erected to the surface of the water at the point A, and from the points *m* and *p*, let the lines *m n* and *p q*, be drawn perpendicular to this line. The angle *n* A *m*, is the *angle of incidence*, and the angle *p* A *q*, is the *angle of refraction*. If now we measure the lines *m n* and *p q*, and divide the length of *m n* by that of *p q*, we will obtain a quotient which is called the *index of refraction.*

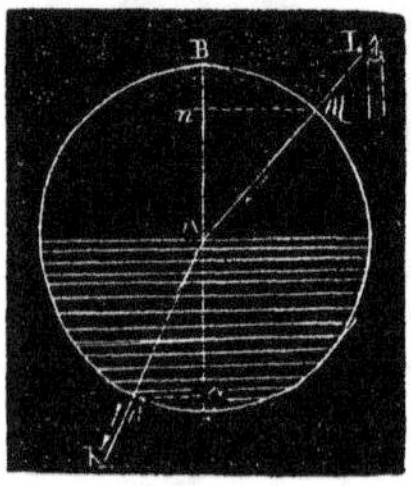

Fig. 95.

It is evident that if the beam were bent still more at

A, this quotient would be larger: the larger the index of refraction, the greater the refracting power of the substance.

For water, the index of refraction is always 1.336; for crown glass, the index is 1.58; for the bisulphide of carbon, it is 1.673.

(70.) The effects of lenses are explained by the principles of refraction.

A convex lens collects the rays of light which pass through it.

A concave lens separates the rays which pass through it.

1. *Lenses.*—A lens is a transparent body bounded by surfaces, one at least of which is curved. Six different varieties are used in the arts. They are usually made of glass, and their shapes are represented by sections in Fig. 96.

The *double convex lens* A, is bounded by two convex surfaces.

The *plano-convex* B, is bounded by surfaces, one of which is convex and the other plane.

Fig. 96.

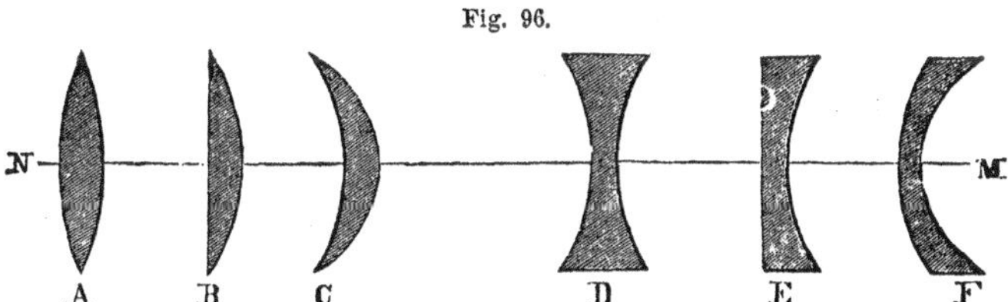

The *meniscus*, C, has one surface convex and the other concave, the convexity being greater than the concavity.

The double concave lens D, has two concave surfaces.

The *plano-concave lens* E, has one surface concave and the other plane.

The *concavo-convex lens* F, has one surface convex and the other concave, the convex surface being less curved than the concave surface.

The first three of these varieties, A, B, C, are *convex* lenses; the others, D, E, F, are *concave* lenses.

2. *The effect of convex lenses.*—By remembering the two laws of refraction, and that a radius of a sphere is always perpendicular to its surface, it will not be difficult to trace the rays of light as they are refracted in going through a lens. Let a section of a double convex lens be represented by M N, in Fig. 97. The two

Fig. 97.

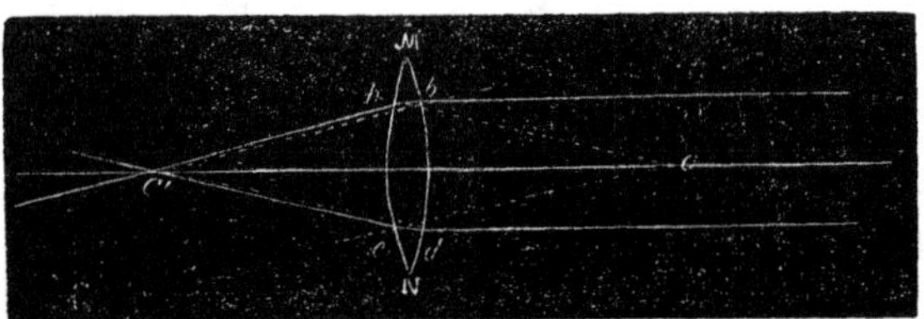

curved surfaces are parts of the surfaces of two spheres, whose centers are at C and C′. The line C C′, drawn through these centers of curvature, is called the *axis* of the lens. Now suppose two rays of light, *a b* and *c d*, parallel to the axis, to fall upon the lens at the points *b* and *d*. On entering the denser medium, they will be bent toward the perpendiculars to the surface at these points. These perpendiculars are the dotted lines C′ *b* and C′ *d*. The refracted rays in the lens go in the directions *b h* and *d e*. On passing out of the lens into air, they are bent from the perpendiculars to the surface at the points *h* and *e*. These perpendiculars are the lines C *h* and C *e*, and the refracted rays cross each other at the point C′. Parallel rays refracted by a convex lens are made converging. And we should find that in all cases, the rays after refraction will be nearer to each other than before. The general effect of the convex lens is to *collect* rays of light.

The plano-convex lens and the meniscus will have the same effect, but in a less degree.

The point C′ is the *principal focus* of the lens: it is the focus of rays which are *parallel* to the axis. The distance of this point from the lens will depend upon the curvature of the lens, and upon the index of refraction. If the two surfaces of the lens are equally curved, and it be made of glass whose index of refraction is 1.5°, then the principal focus will be at the center of curvature, as in Fig. 97.

3. *The effect of concave lenses.*—That concave lenses separate rays of light, may be shown by tracing the rays represented in Fig. 98.

Fig. 98.

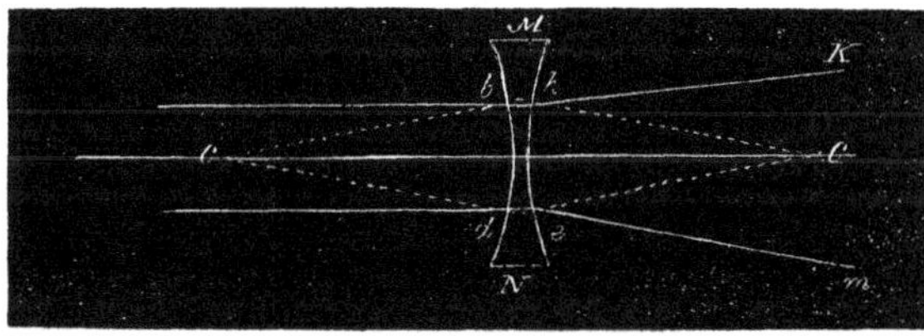

Let M N represent the double concave lens, whose centers of curvature are C and C′. Two rays of light, parallel to the axis, striking the lens at the points *b* and *d*, will be bent toward the perpendiculars, and pass through the glass in the direction *b h* and *d e*. On emerging, they will be bent from the perpendicular, and go in the directions *h k* and *e m*. We thus find that parallel rays are made diverging. Diverging rays would be made more diverging, while converging rays would be made less converging. In all cases, rays refracted by a double concave lens would be *separated.*

The plano-concave lens, and the concavo-convex lens have the same effect, but in a less degree.

(71.) If an object be placed in front of a convex lens, an image of it will be formed on the other side of the lens. To explain this, remember that the image of any point will be made where rays of light going from it, either meet, or appear to meet, after refraction.

1. *Images are formed.*—If a convex spectacle glass is held in front of a window, at some distance, and a sheet of white paper is put in front of it, the light from the window will go through the glass and fall upon the paper. If the distance from the glass to the paper be just right, a very small but very perfect image or picture of the window will be seen upon it.

If a good double convex lens, three or four inches in diameter be at hand, a very beautiful experiment may be made. Through an opening in a shutter of a darkened room, admit a beam of sunlight. Into this beam put any small, transparent object, it may be a picture painted on glass, or, quite as well, a wing of the dragon-fly, so that the light may pass through it. If now, the lens be moved back and forth in front of this object, until just the right distance is found, a very large and perfect image will be seen inverted upon the opposite wall of the room.

2. *The images of points.*—Now let us see how these beautiful effects are produced.

Suppose an arrow N S (Fig. 99), placed at some distance in front of a convex lens M, whose centers of curvature are f and f'. Two rays of light from the point N, passing through the lens, will be refracted so as to cross each other at the point *n*. This point where

Fig. 99.

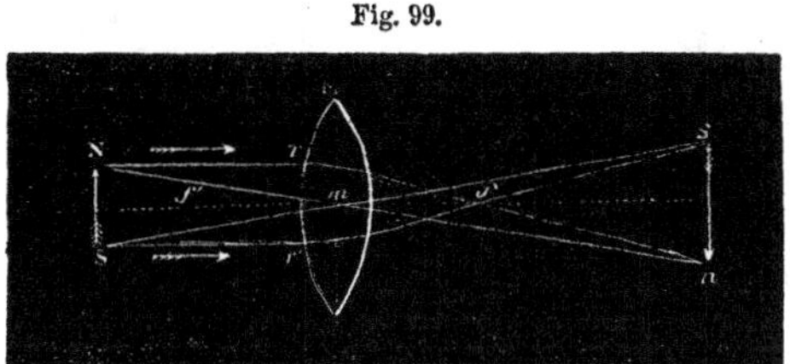

rays of light *meet after refraction, is the image of the point* N, *from which they came.* The rays from the point S of the object, after refraction, cross each other at *s*, and form an image there. From points between N and S, rays of light going through the lens will be collected on corresponding points between *n* and *s*, and thus a perfect image will be made inverted at *n s*.

In this way, it is easy, by a diagram, to illustrate the formation of all images by lenses.

(72.) If an object be placed at a point twice the focal distance from a convex lens, an image of it will be formed at an equal distance on the other side. If the object be moved farther away, or nearer to the lens, the position and size of the image will be changed.

1. *The object twice the focal distance.*—Suppose the lens to be one whose focus is at the center of curvature, and that the object is just twice that distance from the lens, as shown by the arrow N S (Fig. 100). Two rays of light from the top of the arrow go through the lens, bending according to the laws of refraction, and cross each other at the point *n*. Two rays from the bottom

of the arrow go through the lens and cross each other at the point *s*. Join the points *n* and *s*, and *n s* represents the image that is formed. This image will be at

Fig. 100.

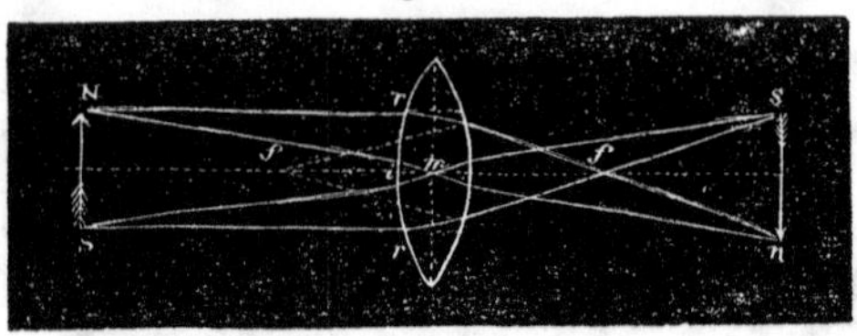

twice the focal distance on the other side of the lens, of the same size as the object, and inverted.

2. *The object farther away.*—This case is represented by Fig. 99. Suppose that in front of the lens M, an arrow, with its head downward, represented by *n s*, is placed at more than twice the focal distance from the lens. Two rays from the arrow-head, after refraction, will be found to cross each other at N; two rays from *s* will, after refraction, cross each other at S. The image N S, is on the other side of the lens, at a less distance, smaller than the object, and inverted.

3. *The object at a less distance.*—If, in Fig. 99, we suppose N S to represent the object, outside the focus, but at less than twice the focal distance, its image will be found at *n s*. In this case the image will be at a greater distance on the other side of the lens, larger than the object, and inverted.

4. *The object between the focus and the lens.*—One more case remains to be considered. Suppose the object to be between the focus and the lens. Let M N (Fig. 101), represent a lens whose focus is at C, and let the object A B, be placed between this point and the lens. An attentive examination of the figures show

that the rays of light from the point A, are diverging after refraction. And since they can never meet, it is clear that no image can be formed on that side of the lens, but if an eye at E, receive these rays they will produce the same effect as if they came from A′. In like manner, the rays from B, entering the eye at E, will seem to have come from B′. Hence an image will seem to be formed at A′ B′. This image will be on the same side of the lens as the object, erect, and larger than the object.

Fig. 101.

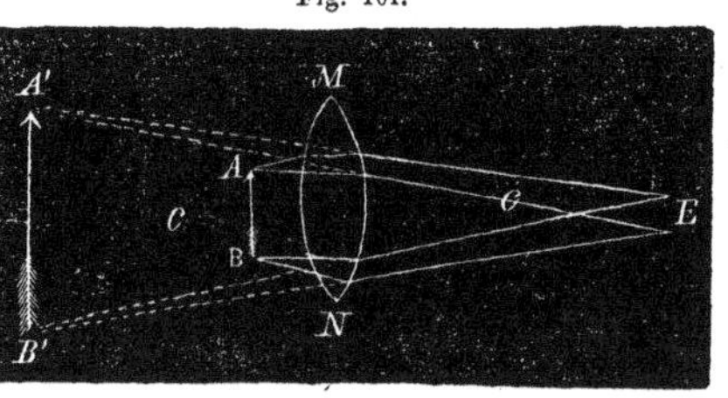

(73.) Images are also formed by concave lenses. They are on the same side as the object, smaller, and erect.

1. Suppose an object A B (Fig. 102), in front of a

Fig. 102.

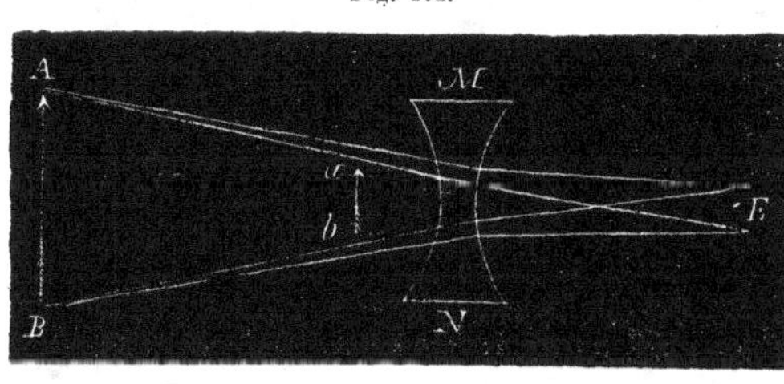

concave lens M N. Rays of light from A, after refraction, diverge as if they had come from *a*; rays from B, after refraction, diverge as if they had come from *b*; the image will thus appear to be made at *a b*. This image is on the same side of the lens, smaller than the object, and erect.

§ 4. ON THE DECOMPOSITION OF LIGHT.

(74.) Prisms refract light; they also decompose it. They separate white light into rays of seven different colors, viz.: violet, indigo, blue, green, yellow, orange, and red.

1. *Prisms.*—Any transparent body, two of whose sides are inclined toward each other, is a *prism.* The most common form of the prism is a triangular piece of glass. A water prism may be made by taking a three-cornered vessel, with glass sides, and filling it with water. Other fluids may be used in place of water.

2. *Prisms refract light.*—Light, in passing through prisms, must obey the laws of refraction. In Fig. 103,

Fig. 103.

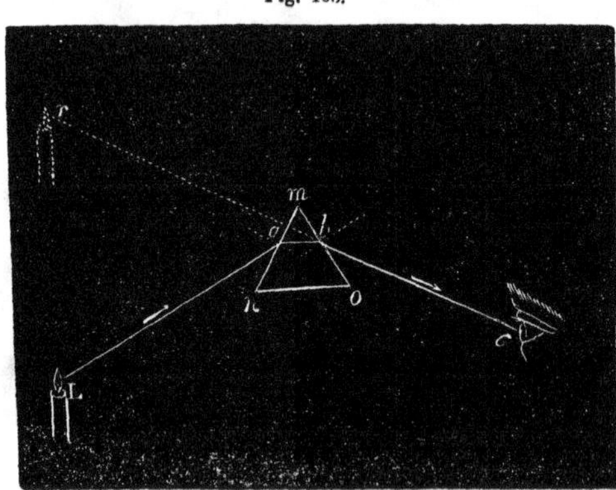

the triangle *m n o*, represents a section of a prism. A ray of light striking its surface at *a*, will be bent toward a perpendicular on entering, and from a perpendicular on emerging, finally taking the direction *b c*. To the eye at *c*, this light would seem to come from the object at *r* instead of L.

3. *Prisms decompose light.*—The white light that comes from the sun, or from other luminous bodies, is really made up of seven different kinds of light. The way in which Sir Isaac Newton made this great discovery is shown in Fig. 104. In the window-shutter S, of a darkened room, he made a small hole, and

Fig. 104.

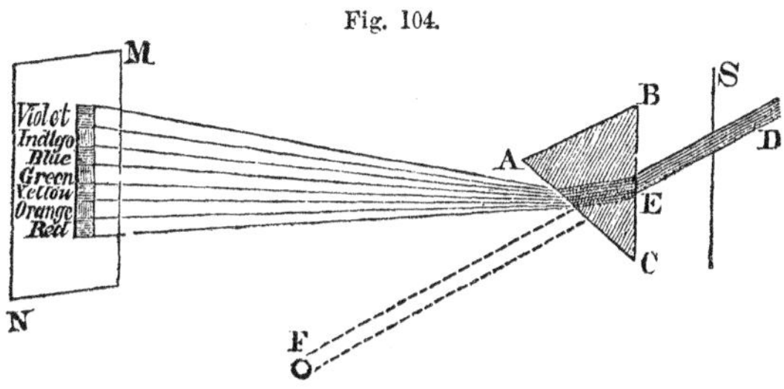

placed behind it a prism, A B C, so that the beam of sunlight D could fall obliquely upon one of its sides at E. Were it not for the prism the beam of light would go straight forward to F, where it would make a round white spot, but being refracted by the prism, it formed above F, upon the screen M, an oblong image containing *seven different colors*. These colors appeared in order from the top of the image, violet, indigo, blue, green, yellow, orange, and red.

These colors are separated, it seems, because the prism has power to bend some of them more than others. The violet rays are bent most; the red rays least.

The oblong image upon the screen is called the *solar spectrum.* The power of a prism to separate the color of white light is called *dispersive power.*

The prism in this way enables us to analyze white light, or to find out the colors of which it is made; and now, if by any means we can unite these seven colors, we shall produce white light again. This can be done by using *any instrument which collects rays of light.* If the rays fall upon a concave mirror, they will be reflected to a focus, which will be a *white* spot. If the rays are received upon a double convex lens, they will be refracted to a focus, and this focus will be also *white.* Sir Isaac Newton collected the rays by using a second prism, exactly like the first, but placed beside it so as to bend the rays in the opposite direction: the image on the screen was *white.*

(75.) The spectrum formed by sunlight or by starlight is crossed by a great many fine black lines, while the spectra formed by light from artificial sources, are crossed by different colored bright lines.

1. *The black lines.*—The whole length of the solar spectrum, when seen by the naked eye, seems to be colored, but when seen through a magnifying glass, a great many fine *black* lines are found to cross it, as if a delicate brush, dipped in the purest black, had been drawn across it by a skillful artist. A beam of sunlight always gives the same set of lines, holding the same relative position in the spectrum. A beam of starlight gives a different set, and the light from different stars gives each a set of its own. These lines are usually called *Fraunhofer's* lines, in honor of him who first examined them carefully.

2. *The bright lines.*—When the light from an artificial source is passed through a prism and its spectrum is seen through a magnifying glass, no black lines are visible, but instead of these, there will be seen lines of exceeding brightness, and of different colors. The color of these lines, and their place in the spectrum, will depend upon the substance whose flame gives the light. If, for example, a little common salt be burned in a hot gas flame, two yellow lines of surprising brightness will *always* appear in the yellow part of the spectrum, while the metal potassium in the flame will *always* give two lines, one of a brilliant crimson color, in the red end of the spectrum, the other a beautiful blue line away off in the violet end. Each substance gives a set peculiar to itself.

(76.) Drops of rain may decompose the sunlight: in this way the rainbow is produced. The primary bow consists of bands of the seven colors of the spectrum, arranged in parallel arches, with the red band on the outside.

In the secondary bow the order of the colored arches is changed, the violet being on the outside.

Fig. 105.

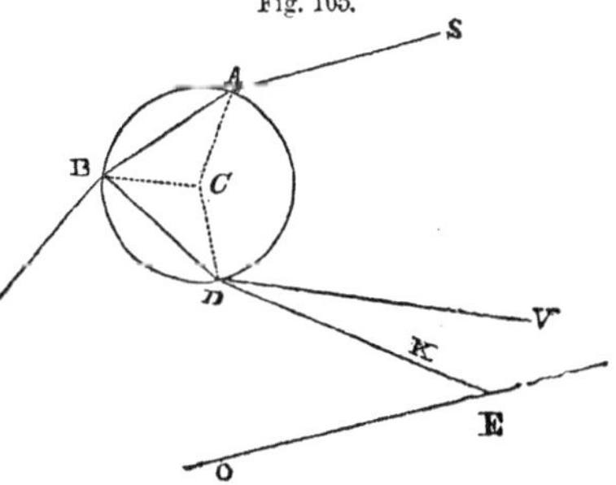

1. *The primary rainbow.*—This most beautiful phenomenon is produced by the action of rain drops; they decompose the sunlight and send its rich colors to the eye.

To understand this action, suppose the

circle, whose center is at C (Fig. 105), to represent a section of a drop of water. Rays of sunlight (S A) falling upon the upper part of the drop will be refracted to the point B. At this point a part of the light will pass out into the air again, but another part will be reflected by the inner surface of the water and strike the surface at another point, D. The light which here goes out of the drop into the air, will be again refracted. Now the light will not only be refracted, in its passage through the drop; it will be, at the same time *decomposed.* On coming out of the water the red ray, bent least, will take a direction represented by D E; the violet ray, bent most, may be represented by D V; and all the other colors of the spectrum will be found between these.

2. *The red band on the outside.*—Now it is quite clear that if the person were standing upon the ground in the direction of D E, so that the red rays from this drop would enter his eye, the violet rays, and indeed all the other colors, would go over his head. To him this drop of water would appear red. Another drop, some distance *below* this one, would send violet rays into the same eye. Between the drop which sends the red, and that which sends the violet, there would be others from which the eye would receive the other colors of the spectrum. (Fig. 106.)

Fig. 106.

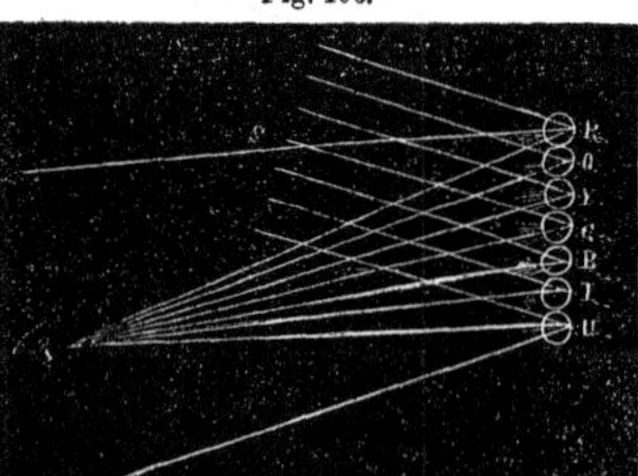

Hence, when a shower of rain is falling, and the sun is at the same time shining in the opposite part of the sky, so that a person looking toward the shower, will have his back turned toward the sun, he will see the seven colors of the spectrum painted upon the cloud in order, with red at the top and violet at the bottom.

3. *The colors are in the form of an arch.*—Now, suppose a line drawn from the sun through the eye of the observer, and straight onward until it reaches a point O (Fig. 105), directly under the drop C, which sends the light to the eye. If this drop sends a red ray to the eye, then all others, which like this are opposite the sun, and whose distance from O is the same, will also give red rays. If the arc of a circumference be drawn with O as a center, and with a radius C O, all drops along this circumference will be equally distant from the center O, and will therefore give red rays. The red part of the rainbow is, for this reason, a circular arch, and for the same reason, the other colors are parallel arches below the red.

4. *The secondary bow.*—Outside of the bow just explained, another, the secondary bow is often seen. Its colors are more dim, and their order is reversed, the violet being at the top and the red at the bottom.

To explain the primary bow we trace rays of light falling upon the top of the drops of water. But drops of rain in the air are entirely covered with light, and to explain the secondary bow, we may trace the rays which fall upon their lower parts. The diagram (Fig. 107) illustrates this. A beam of light S A, refracted on entering the drop, goes through to its

Fig. 107.

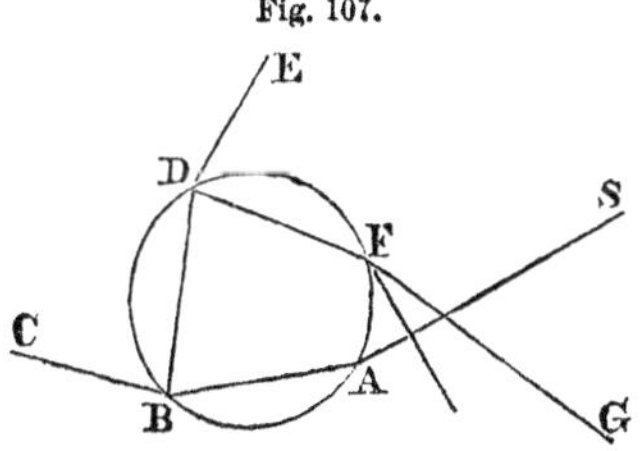

inner surface at B, from which it is reflected. It strikes the inner surface again at D, and is again reflected. At the point F, a part of the beam will be again reflected, but another part will pass out into the air and be bent downward. In its passage through the drop the light is not only refracted; it is decomposed. If F G represents the red ray, then F V may represent the violet ray. Now, clearly, if the red ray enters the eye, the other colors will fall *below* it, so that this drop will appear red. Other drops above this one will give the other colors in their order. Hence the outside band of the secondary bow will be violet, the inner one red.

(77.) Bodies are of different colors, only because they decompose the sunlight, and reflect different parts of it to the eye. The various colors of the sky, and the clouds, are due to the decomposition of the light which comes through them from the same.

1. *The color of bodies.*—The sun sheds a flood of pure white light upon all bodies alike. This white light is decomposed at their surfaces. Some of its colors are transmitted or absorbed by the body, while the others are reflected to the eye. One body is red because it decomposes the sunlight and reflects the red rays; another is blue, because it reflects only blue rays. The foliage of trees in the spring-time, receives the sun's white light, decomposes it, and reflects only the green rays. The petals of the violet decompose the sunlight to share with us the beautiful colors of the spectrum; it reflects the colors of the violet end, and keeps to itself those of the other. A body which

reflects all the color of the light it receives is white; one which reflects none is black.

2. *The color of the sky.*—The sky, when free from clouds, is blue, because the particles of the atmosphere reflect blue rays of light. If the thin air could not reflect light at all, the sky would appear black: if it reflected it without decomposition it would be white. The white sunlight falls upon its molecules, is decomposed by them, and only those rays which make up the delicate blue color of the sky are reflected to our eyes.

3. *The color of the clouds.*—The clouds both reflect and refract the sunlight, and all their varied colors are due to the decomposition thus produced. There can be no more gorgeous display of colors than we often see upon the clouds of the morning and the evening sky. What grand and diversified effects to be produced by means of such simple materials as light, water, and air!

§ 5. ON OPTICAL INSTRUMENTS.

(78.) The microscope, the telescope, and many other instruments, help the eye to see small or distant objects, by forming large and perfect images of them near by, for it to examine.

The eye itself is an optical instrument of the most perfect construction.

1. *The microscope.*—The simple microscope consists of a single convex lens. The lens is held in the hand at a little less than its focal distance from the object. The eye receives the light which comes from the object through the glass, and sees a magnified image on the other side.

The operation of the *compound microscope* may be

understood by means of a diagram (Fig. 108). Two convex lenses, and sometimes three, are used.

Fig. 108.

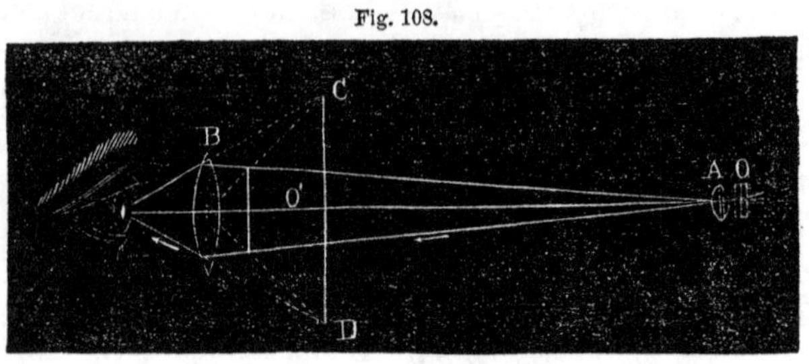

The lens A, called the *object-glass*, refracts the light from the object O, placed a little beyond its focus, and forms an image inverted at O′. The light from this image is refracted by another lens B, called the *eye-glass*, and if the rays are received into the eye, they will appear to have come from C D, which is the magnified image of the object.

By means of this instrument, things otherwise too small to be seen, are made visible, and a world of wonderful creations is thus revealed for the study and admiration of man. A drop of water from a stagnant pool, is found, by means of the microscope, to be swarming with living creatures, whose forms are as perfect, and whose appetites are not unlike those of larger animals.

2. *The telescope.*—A telescope is used for viewing distant objects. Sometimes a lens is employed to form an image; sometimes the image is formed by a mirror. In the first case, the instrument is called a *refracting telescope;* in the second, it is called a *reflecting telescope.*

Of the refracting telescope there are three important forms: *Galileo's*, the *astronomical*, and the *terrestrial.*

In *Galileo's* telescope there is a double convex object-glass M N (Fig. 109), and a double concave eye-glass,

Fig. 109.

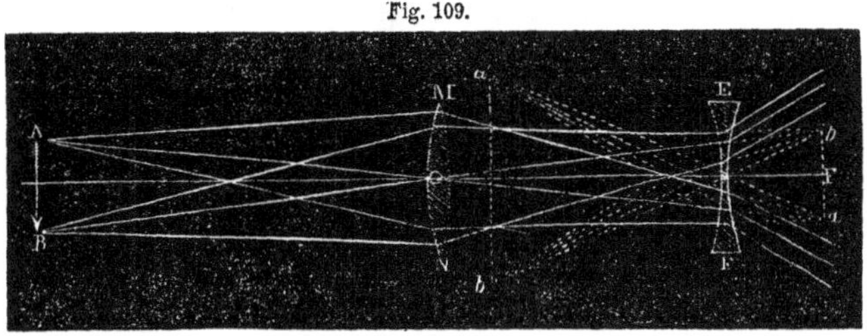

E F. Rays of light from the point A of a distant object A B, after passing through the two glasses, diverge as if they came from the point *a*, while rays from the point B of the object after refraction, diverge as if from the point *b*. An erect image *a b*, will be seen by holding the eye in front of the eye-glass E.

The *opera-glass* consists of two small galilean telescopes placed side by side.

In the *astronomical* telescope, two double convex lenses are used. The object-glass O (Fig. 110), forms a

Fig. 110.

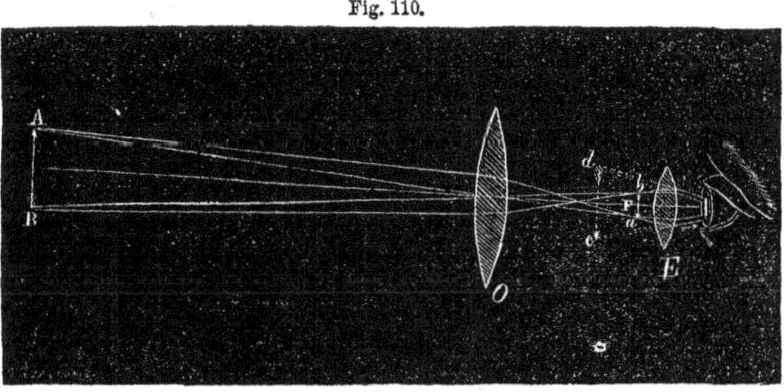

small image *a b* of a distant object A B. The eye-glass (E) being so placed that its focus (F) is a little beyond this image, refracts the light, so that it will appear to have come from a magnified image *c d.* The course of the rays may be traced in the figure. In this instrument the image is always inverted.

The *terrestrial* telescope is used for viewing distant objects upon the earth. To see them upside down, as in the astronomical telescope, is not desirable: that they may be seen right side up, two convex lenses are placed between the object-glass and the eye-glass. The arrangement of the glasses, and the

Fig. 111.

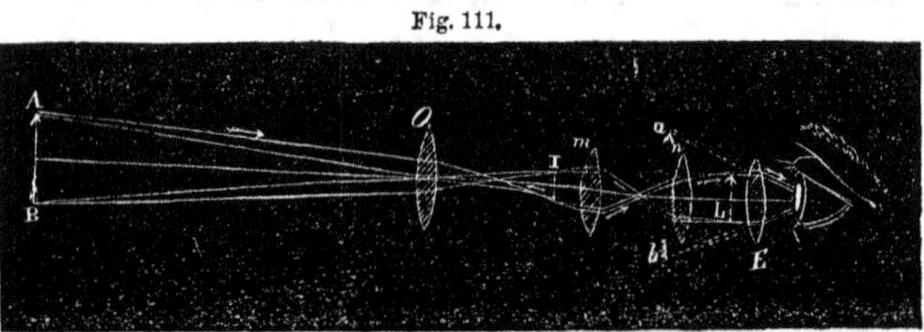

course of the rays, are shown in Fig. 111. The object-glass O, forms a small inverted image I, of a distant object A B, near its focus. From this image the light goes through the two lenses, *m* and *n*, to form a second image L. This image is *erect* with respect to the object, and it is magnified by the eye-glass E, in the usual manner.

Of the reflecting telescope there are several varieties. In all of them the image of a distant object is formed by a concave mirror, and this image is magnified by a convex eye-glass.

In the *Herschelian telescope* (Fig. 112), the mirror M N is inclined to the axis of the tube in which it is placed, so that rays of light from a distant object will be reflected to a focus near to one side of the tube at the other end. The observer, looking down into the tube, holds an eye-piece, *a*, in his hand, through which he views a magnified image.

Fig. 112.

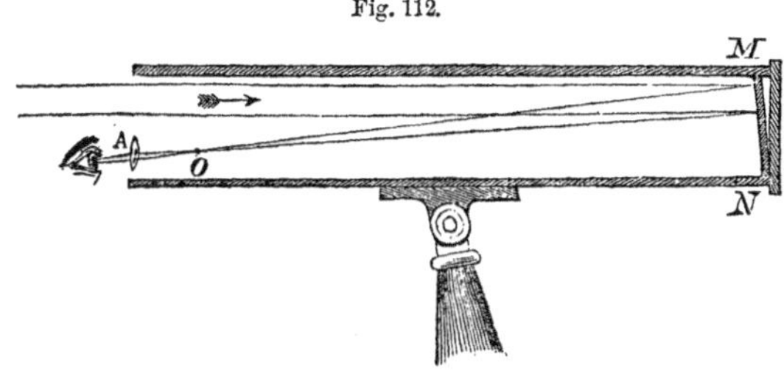

3. *The magic lantern.*—The magic lantern is an instrument by which the image of a small transparent picture, painted on glass, may be thrown upon a screen, so much magnified that a whole audience may see it.

It consists of a powerful lens, with objects highly illuminated by lamp-light placed so near it, that their images are formed far away. Fig. 113 shows a section of the instrument. Inside of a dark box, a strong

Fig. 113.

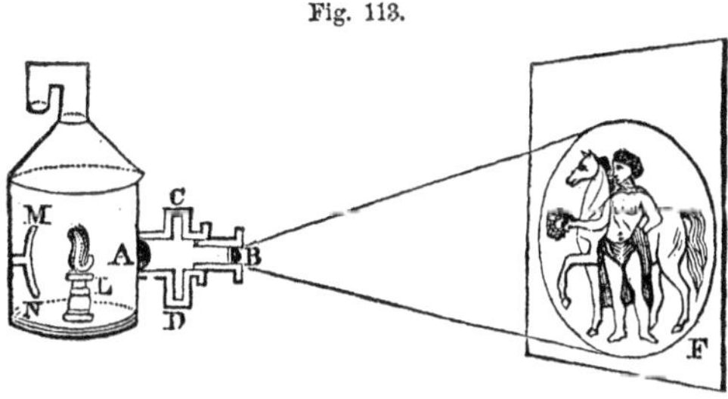

light (L) is placed. Behind this light is a concave mirror (M) and in front of it a convex lens A. This lens is at the entrance of a tube which projects from the side of the box. Inside this tube slides a smaller one, in which is fixed another powerful lens. The picture is placed in a slit C, provided for it in the larger tube, just in front of the first lens. The lamp fills the box with a strong light. The lens A, receiving light directly from the lamp, and reflected from the mirror, condenses it upon the object and highly illuminates it. The light from this bright object goes on through the second lens to the distant screen, and there forms a large and perfect image.

This instrument is very useful to the lecturer or the teacher, who would illustrate the wonderful phenomena of nature. By means of small pictures, or of small transparent objects, he is able to make his audience *see* the relations of the heavenly bodies taught in Astronomy, or the delicate phenomena described in Natural Philosophy and Chemistry.

4. *The camera obscura.*—The camera obscura is an instrument by which to form miniature images of objects. It consists of a dark box, a section of which is represented by A B (Fig. 114), containing a screen S,

Fig. 114.

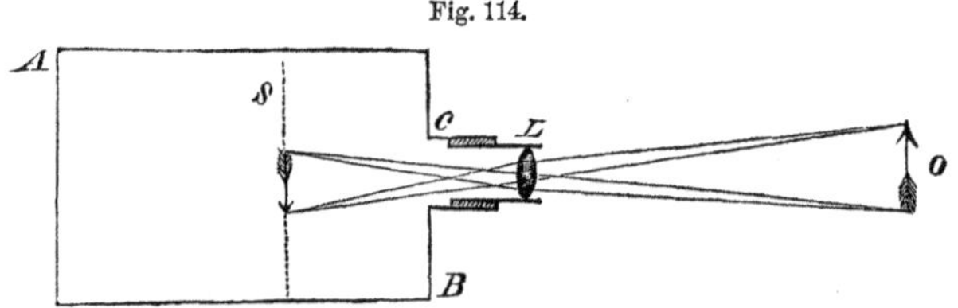

and having a double convex lens L, filling an opening in one end. The distance of the lens from the screen may be varied by sliding the tube which carries it back and forth in the larger tube C. The light from the object O is refracted by the lens, and a beautiful image will be formed upon the screen. This image is always inverted, and smaller than the object.

The camera may be illustrated by a very simple experiment. If, in a hole in the shutter of a darkened room, is placed a double convex lens, the room is itself a camera obscura, and persons present may see what takes place inside. Let a white sheet be hung in front of the lens at a proper distance, and it is at once covered with a perfect picture of whatever scenery may be outside. Houses and distant hills; the sky with its floating clouds; men and animals in the street, and even the flying birds, and the curling smoke, are painted upon the screen, with colors taken from the sun's bright rays.

5. *The eye.*—But most perfect of all optical instruments, is the eye. Who could at first believe, that in describing, as we have done, the camera obscura, we were describing a rough model of the human eye! Yet the eye is nothing but a simple camera obscura, differing from it only in its wonderful perfection.

The human eye is a globular chamber, having for its outer wall a hard tough membrane called the *sclerotic coat.* The front part of the sclerotic coat is a transparent substance called the *cornea.* The chamber is lined with a more delicate membrane called the *choroid*, and to insure the darkness of the place, this is covered upon the inside with a *black paint.* The front part of the choroid coat is called the *iris*, and in the center of this is a round hole called the *pupil* of the eye, through which light may pass into the dark chamber beyond. Behind this opening, is a double convex lens, very transparent and considerably hard, called the *crystalline lens.* Between this lens and the cornea is a limpid liquid called the *aqueous humor*, and filling the dark chamber, behind the lens, is another fluid, called the *vitreous humor*. The arrangment of these parts may be understood by attentively studying Fig. 115, which

Fig. 115.

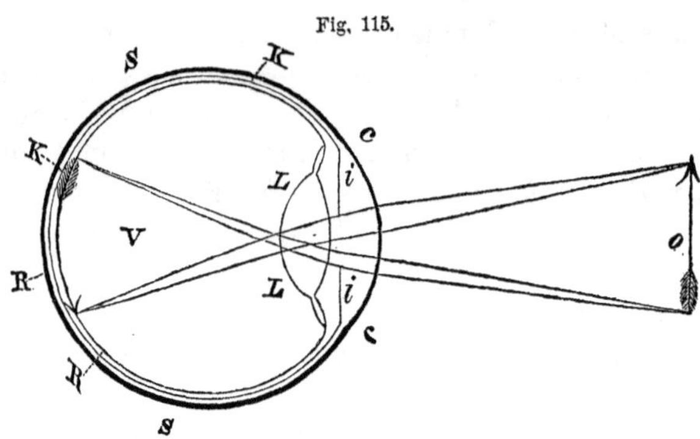

represents a section of the eye. S S, is the outer or sclerotic coat, sometimes called the white of the eye. C C, is the cornea; it is more convex than the sclerotic. K K, is the choroid, and *i i*, is the iris, the vertical curtain which shuts out all light, except what may get through the hole at its center—the pupil. L L, is the crystalline lens, and the large chamber V, is filled with the vitreous humor. The course of the rays of light is also shown in the figure. An inverted image of an object O, is formed at R. It is there received upon a net-work of delicate nerve fibers called the *retina*, R R. The mind takes cognizance of this picture, and the person is said to *see the object* O. These pictures on the retina are always smaller than the objects, and the more distant the object, the more minute the image. The diameter of the eye is little more than an inch, and yet when a person sees an extended landscape, every visible object, far and near, is painted upon the inner lining. If the picture in the human eye be thus minute, what must it be in the eye of a canary-bird or butterfly!

CHAPTER VII.

THE EFFECTS OF VIBRATIONS.—III. HEAT.

§ 1. OF THE SOURCES AND NATURE OF HEAT.

(79.) THE sources of heat may be studied in three groups. First, the heavenly bodies; second, mechanical action; and third, chemical action.

I.—THE HEAVENLY BODIES.

A.—The sun and the stars are sources of heat.

1. *The sun.*—Floods of heat come down with the sunlight. Upon this most familiar fact we need not dwell, further than to notice that the amount of heat received from the sun is doubtless greater than from any other source, and that as this amount varies from month to month, it allows the earth to be clothed in the snows of winter, the verdure of spring, the maturing growths of summer, and the ripening fruits of autumn.

2. *The stars.*—Heat comes with the starlight as well as with the sunlight. During the night when the stars are seen, not more than during the day when the stronger light of the sun obscures them, each star is sending its proportion of heat to warm the earth.

II.—MECHANICAL ACTION.

B.—No mechanical action can occur without evolving heat in the bodies which act upon each other.

1. *Mechanical action evolves heat.*—When the savage lights his fire by rubbing two pieces of hard wood together, he produces heat by *friction.* When by repeated blows of a hammer, a nail is made too hot to handle, or when the iron-clad hoof of a horse "strikes fire" against a pavement stone, heat is evolved by *percussion.* And finally, when a piece of cold wood is heated by being squeezed between the plates of a hydrostatic press, heat is evolved by *pressure.* No two bodies can act upon each other, either by friction, by blows, or by sudden pressure without evolving heat.

2. *Friction.*—By friction heat may be evolved in *large quantities.* Count Rumford caused $18\frac{3}{4}$ lbs. of water—almost two gallons, to boil by the friction of a solid plunger against the bottom of an iron cylinder immersed in the fluid. *All bodies*, whether solid, liquid, or gaseous, give off heat by friction. Sir Humphrey Davy quickly melted two pieces of ice by simply rubbing them together in a room whose temperature was below the freezing point. A bullet is warmed by the friction of the air through which it passes. A stream of water is warmed by friction against the sides of a channel through which it swiftly runs. Moreover, the production of heat by friction is *unlimited;* it will continue just as long as the friction is kept up.

III.—CHEMICAL ACTION.

C.—*Chemical action a source of heat.*—Combustion

is the most familiar form of chemical action ; it is at the same time the most common source of artificial heat. Wood burns in the stove, or coal in the grate, and our houses are warmed by the heat given off by this chemical action. A chemical action takes place in the body of a person by which the food is changed to blood, and the blood again to bone and muscle: heat is evolved by this chemical action, which keeps up the constant temperature of the body. It is thought that no chemical action can take place without producing heat.

(80.) The material theory of heat supposes it to be a very subtile fluid which fills the spaces between the molecules of bodies, and whose presence in larger or smaller quantities constitutes heat or cold.

The dynamic theory supposes that heat is the result of vibrations among the molecules of a body : the more rapid the vibration, the higher the temperature.

1. *The material theory.*—The material theory supposes matter to consist of molecules, that these molecules do not touch each other, and that the space between them is filled by a substance called *caloric*, whose molecules are very much smaller than those of the body. Just as when water is poured into a barrel already filled with bullets, it runs into the spaces between them, so it is thought that the fluid caloric goes into and fills the molecular spaces.

The sun and stars throw off abundance of this substance, and shoot it with the velocity of light across the spaces between them and the earth. In the cases of friction, of blows, and of pressure, the molecules of bodies are pushed nearer together, and by this means the caloric is *squeezed* out from between them, as, when a wet sponge is pressed, water flows from its interstices. In chemical action also, the particles of bodies are brought nearer together and force the heat fluid from between them. This theory, which until lately was opposed by only a few eminent men, is now almost universally discarded.

2. *The dynamic theory.*—It is now generally believed that heat is the result of vibrations. This theory, like the other, supposes matter to be made up of molecules separated by definite distances; it goes further, and supposes these molecules to be *in motion*, rapidly vibrating in the minute spaces between them. To increase the rapidity of this motion is to make a body hot; to lessen it is to make the body cold. The theory assumes also the existence of the *ether*, which according to the theory of light must fill all space.

When we step from the shade into the sunlight, the gentle heat of its rays is instantly felt. The explanation is this: the molecules of the sun itself are in rapid vibration, they impart motion to the ether, whose vibrations dart through the space between the sun and us, and, coming in contact with the person, impart vibration to the molecules of the sense of touch, when we become immediately conscious of the presence of heat.

When bodies are heated by friction, their molecules are made to vibrate faster by the rubbing. Heat is evolved by percussion, because a blow increases the motion of the already trembling particles of the body struck. The same effect is produced by pressure.

§ 2. OF THE TRANSMISSION OF HEAT.

(81.) Rays of heat, like rays of light, pass through

some bodies more freely than through others. They obey the same laws of transmission, of reflection, and of refraction.

1. *Rays of heat.*—Since heat and light come together in the sunbeam, and since they are thought to be of the same nature, both being the result of vibrations, we may speak of *rays* of heat, just as we do of *rays* of light.

2. *Transmission of heat rays.*—Just as light passes more freely through some bodies than through others, so heat passes through different bodies with different degrees of facility. Those bodies through which it passes most freely, are said to be *diathermic*, while those through which it can go with the greatest difficulty, are said to be *athermic.*

Heat from different sources is transmitted in different degrees through the same substance. It is, for example, a familiar fact that the glass of our windows allows the heat of the sun to enter our rooms, while it prevents the heat of the stove from going out.

Rock-salt is the most diathermic substance known; it allows heat from all sources to pass through it with the greatest freedom.

3. *Laws of transmission.*—Heat passes through space with the same remarkable velocity as light. It obeys the same laws of transmission. [See (60).]

4. *Law of reflection.*—Heat is also reflected in the same way that light is. For the law which it obeys, see (61).

In former times, when the open fireplace was common, the housewife baked her bread by heat reflected from the top of a tin oven placed before the fire. This oven, once found in every kitchen, now only in the garret if found at all, having been pushed out by the modern stove, consisted of a tin box closed at the back and the ends, open in front, and having its top slanting at an angle of about 45°. A horizontal shelf was placed in the middle of the oven, upon which stood the loaf to be baked. Under the shelf was another slanting tin surface. The oven standing with its open face to the blazing fire, received the heat rays upon its two slanting surfaces, and reflected them against the top and bottom of the loaf.

5. *Law of refraction.*—Heat is also refracted like light. [See (68).]

The sun's heat coming through the window-glass is bent from its course. By a double convex lens it may be collected, and its intensity greatly increased. The common burning glass illustrates this. It may be a spectacle glass held by the hand in a sunbeam, and the small bright spot—the focus of light—is also the focus of heat. The other hand held at this point will be burned; tinder will be set on fire, or gunpowder exploded.

(82.) Heat tends to diffuse itself equally among all bodies. This distribution takes place in three ways; by conduction, by convection, and by radiation.

1. *The equal diffusion of heat.*—If two bodies, one cold, the other hot, be placed near each other, it will in a short time be found that both are equally warm. The cold body has received more heat, the hot body has parted with some that it had. What is thus true of *two* bodies is true of *all*. Bodies are constantly giving and receiving heat. Those which part with more than they receive from others, get colder; those which receive

more than they give, get warmer. Ice, for example, is giving heat to all bodies around it; it is at the same time receiving heat from them in return. Ice will actually warm a body which is colder than itself, because it will give more heat than it gets in return; it will be melted by a body warmer than itself, because it receives more than it gives.

I.—CONDUCTION.

A.—Heat is conducted through some bodies much more freely than through others. Among solids the metals are the best conductors. Liquids are poor conductors, and gases still poorer.

1. *Conduction.*—Heat is transmitted by conduction when it goes to different parts of the same body by traveling step by step from molecule to molecule.

To illustrate this definition, suppose that one end of a cold iron rod is held in the flame of a lamp. The heat will travel gradually from the flame through the rod, until the distant end gets too warm to be held by the hand.

Now, if we would understand how the heat has made its little journey through the rod, we must picture to ourselves the delicate motion of the molecules of the iron. Those molecules in contact with the flame are made to vibrate; they swing against their neighbors and put them also in more rapid motion; they, in turn, give motion to the next, and these to the next, until those at the distant end of the rod have finally received the shock. The vibrations of these molecules of the rod, impart motion to the molecules of the hand in contact with them; the delicate nerves of touch receive the impulses, and announce the pain. The hand is burned; the rod is hot.

Some bodies conduct heat more freely than others. Those which conduct heat freely are called *conductors:* those which hinder its passage much, are called *poor conductors*, and those which nearly or quite forbid its passage, are called *non-conductors.*

2. *Metals are good conductors.*—Among solid bodies the metals, as a class, are the best conductors, but among metals there is great difference in conducting power. By a very simple experiment this may be illustrated. Plunge two spoons, one of silver and the other of German silver, into the same cup of hot tea; it will be found that the upper end of the silver spoon will get hot much quicker than that of the other. Among the best conductors we find silver, copper, gold, brass, tin, and iron, in the order named.

3. *Conduction in liquids.*—The conducting power of liquids is very feeble. Water, for example, may be boiled in a glass tube, with ice at the bottom without melting it, by applying the heat to the top of the water, or near the upper end of the tube.

4. *Conduction of gases.*—Whether gases conduct heat in the least degree is doubted. Dry air is surely among the poorest conductors, and so, likewise, are all porous substances in which large quantities of air are inclosed.

II.—CONVECTION.

B.—Convection takes place in bodies whose particles are free to move. Air is heated in no other way. Liquids are also heated by convection, but it can not occur in solids.

1. *Convection.*—Heat is transmitted by convection when it is carried from place to place by moving particles of matter. The following very simple experiment will make this definition clear. Upon a plate of thick glass or a smooth block of wood put a bit of candle, lighted, and over it place a lamp-chimney so that its edge may project a little beyond the edge of the block (Fig. 116.) If the edge of the chimney fits closely upon the top of the block so that no air can enter, except at the open part A, the flame will flutter violently, showing that air is forced against it. If now, some light substance, such as down or cotton, be hung from a thread *above* the top of the chimney, it will be lifted away, showing that air is rising out of the chimney. Now, we know already that air is expanded by the heat, and we learn from this experiment that the cold air going under the glass pushes the expanded air away from the flame, up and out at the top of the chimney. This motion of heated air is convection.

Fig. 116.

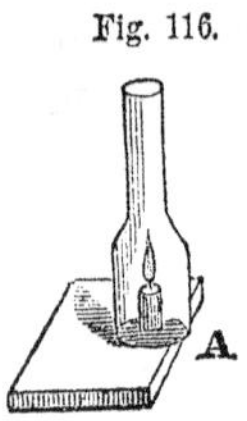

What we have seen in this experiment really takes place whenever a hot body is surrounded by colder air. The air in contact with a hot stove, for example, is heated and expanded. The colder air then pushes it away and takes its place, only in turn to be heated and pushed away by other colder portions. The air goes to the stove, becomes heated, and *moves* away to other parts of the room, *carrying the heat with it.* This transfer of heat by the moving particles of air is called convection.

2. *Air is heated in no other way.*—Air is heated only by convection. The heat of a stove does not go out to

distant parts of a room to warm the air: the air must go to the stove to get warm. So, too, the atmosphere is warmed by convection. The sunbeams coming through it do not warm it; they only warm the earth beneath it. Nor does the heat of the earth pass from particle to particle, as it may in solid bodies; the heat of the earth warms only those particles in immediate contact with it. These rise and carry their heat with them to upper regions, while colder ones take their places in contact with the ground to get warm in turn, and then ascend.

3. *Liquids heated by convection.*—Liquids are also heated by convection. A simple experiment will illustrate the convection of water. Into a flask or bottle of water put a little cochineal. Its particles are just about as heavy as those of the water, and will show by their motion whether there are currents in the water. Warm the bottom of the bottle, and the heated water will be seen to be rapidly leaving the bottom, while other portions are moving downward to take its place.

4. *No convection in solids.*—Solids can not be heated by convection simply because their particles are not free to move among themselves.

III.—RADIATION.

C.—Heat is distributed by radiation from all bodies. The amount of heat which a body can radiate depends upon its temperature, its nature, and the condition of its surface.

1. *Radiation.*—Heat is transmitted by radiation when it goes in straight lines, in all directions, through non-conducting media.

The air is a non-conductor, yet heat passes through it with the greatest freedom. Let a red hot iron ball be suddenly placed in a cold room; a person at a little distance will almost instantly feel its heat rays falling upon his face. The heat is not brought to him by convection, because he feels it just as well whether the ball is *above* his face, below it, or beside it, while the heated air can move only *upward*. It is transmitted by radiation. The heat of the sun comes to us through space where there is no solid body to conduct it, no liquid nor gaseous substance to bring it to us by convection. It is *radiated* to us.

It is chiefly by radiation that the general distribution of heat is accomplished. All bodies, at all temperatures, are radiating heat. That which radiates more than it receives from others, grows cold, while that which gives less than it gets, grows warm.

2. *Depends on temperature, nature, condition.*—The higher the temperature of a body, the more heat it can radiate. Moreover, bodies of different substance, when at the same temperature, give off different quantities of heat in the same time. Thus, iron is a better radiator than gold. Still farther, the same body, at the same temperature, with a *rough* surface will radiate much faster than when its surface is smooth. The rough surface of a cast iron stove, for example, is a better radiator than if it were polished.

§ 3. ON THE EFFECTS OF HEAT.

(83.) The action of heat is twofold; it raises the temperature of a body to which it is applied, and at the same time expands it. This is true of its action upon solid, liquid, and gaseous bodies.

Temperature is measured by the expansion which accompanies it, in instruments called thermometers. There are three varieties of thermometers in use, the Fahrenheit, the Centigrade and the Reaumer. The air thermometer is used to show delicate changes of temperature.

1. *The action of heat is twofold.*—That the temperature of bodies is raised by the application of heat is too familiar to need illustration. That while the temperature rises, the body grows larger, is known by such facts as the following:—A ball of metal, which, when cold, just fits a ring, will be too large to enter it when hot. A clock pendulum is longer in summer than in winter. The tire of a carriage wheel is put on while hot; on cooling, it contracts and binds the parts of the wheel firmly together.

2. *The expansion of solids.*—To show the expansion of a metallic bar, the following beautiful experiment has been devised. A rod of metal (R, Fig. 117), is fixed at one end E, while the other end, passing freely through a post, presses against a bar of brass B. One end of this bar is fastened by a hinge *h*, while upon the other end above, rests another bar D. This second bar also turns upon a hinge, and upon its other end, it carries a small plain mirror M.

If a beam of sunlight S, coming through a hole in the shutter of a partially darkened room, is made to fall upon the mirror, it will be reflected, and form a white spot upon a distant wall or ceiling. This spot will be quite still as long as the mirror does not move, but when the mirror rises, the spot will move along the ceiling rapidly. Now, let the rod R, be heated, and the spot moves. The rod must be expanded, so as to push against the first bar B, the upper end of which pushes against the second bar D, and *lifts the mirror.*

Fig. 117.

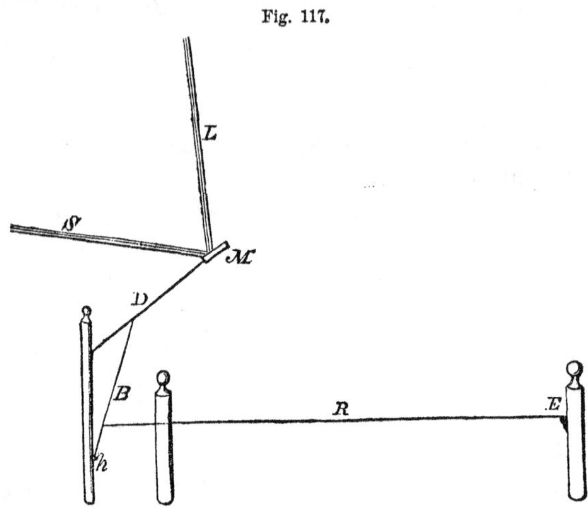

If the air in the room is dusty, the entire beam of reflected light L, can be seen, tracing a luminous path upon the ceiling, while the bar expands.

Different solids expand unequally for the same increase of temperature, but each solid expands uniformly, —that is to say, two, three, or ten degrees of heat will produce respectively two, three, or ten times as much expansion as one degree.

3. *The expansion of liquids.*—The expansion of a liquid by heat, may be shown by filling a bottle with water, and then, after fitting into the cork a glass tube open at both ends, and tightly pressing the cork into the neck, standing it in a vessel of warm water. The water in the bottle will be heated, and its expansion

will be shown by the fluid rising higher and higher in the tube.

A little artifice enables one to change this slow motion into another so rapid, that it may be easily seen even by those in the most distant part of the room. A very long slender body, a rye straw full length answers the purpose admirably, is suspended by a thread tied near to one end. From this end (E, Fig. 118), is hung

Fig. 118.

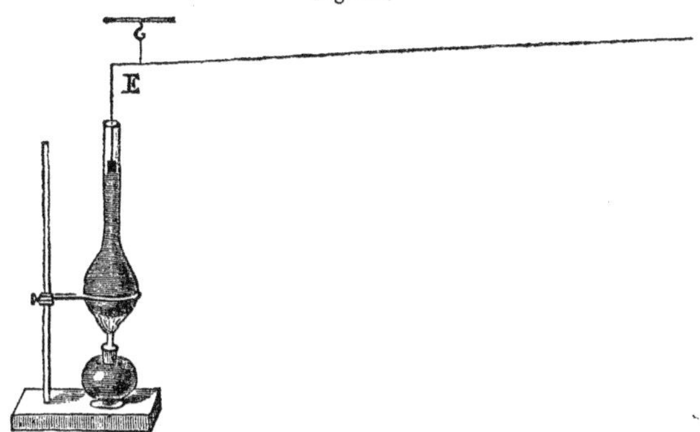

a little cylinder of metal, of such weight as to be almost balanced by the long arm of the straw lever. The cylinder hangs inside of the tube of the bottle, and rests upon the water in it. Now, when the water in the bottle expands, it rises in the tube, and pushes the little cylinder up before it. While it goes up, the other end of the straw lever goes down many times faster. The rapid sinking of the straw, while the water is being heated, shows that the water is expanding.

All liquids are expanded by heat, but some are much more affected than others. The expansion of liquids is much less uniform than that of solids.

4. *The expansion of gases.*—All gases are expanded

by heat. For illustration by experiment, of the expansion of air read (16.) 2.

The expansion of gases is more nearly uniform than that of either solids or liquids, and much greater for the same addition of heat. What is more remarkable is the fact that they *all expand at the same rate.* If we have 490 cubic inches of air at a temperature of 32° F. and add one degree of heat, there will be 491 cubic inches: it expands $\frac{1}{490}$ of its bulk. All gases expand at the same rate, $\frac{1}{490}$ of their bulk, at 32° for every additional degree. This fraction ($\frac{1}{490}$) is called the coefficient of expansion for gases.

5. *Temperature measured by expansion.*—We have found that temperature and expansion increase at the same time by the addition of heat. Moreover, in the same body a certain amount of expansion *always* occurs with the same increase of temperature. By seeing the expansion we may therefore judge of the increase of temperature.

The expansion of solids is too slight, while that of gases is too great, to be conveniently used to measure the changing temperature of the air and other things. *Mercury* is a liquid metal, whose expansion is remarkably uniform, and neither too great nor too little for practical purposes. All common thermometers are made with it.

6. *The thermometer.*—The mercurial thermometer consists of a glass tube terminating at one end in a bulb, and sealed at the other. The bulb and lower part of the tube are filled with mercury, the space in the tube above the mercury being a vacuum. Behind the tube is a graduated scale to show the height of the column of mercury.

7. *Various forms.*—There are three modes of graduating the scale, and this gives rise to three varieties of mercurial thermometer. In *Fahrenheit's* instrument, the zero of the scale marks the height of the mercury in the tube when the bulb is placed in a mixture of snow and salt. When the bulb is put into boiling water, the mercury in the tube runs up to a point which is marked 212 on the scale. The distance between these points is divided into 212 equal parts called degrees, and this graduation is carried above and below these points. According to this thermometer, water boils at 212° and freezes at 32°.

In the *centigrade* thermometer, the zero point marks the height of the mercury in the tube when the bulb is placed in freezing water. The height to which it rises when the bulb is put into boiling water, is marked 100, and the distance between these points is divided into 100 equal parts. The boiling point of water is, therefore, 100° ; its freezing point is 0°.

In *Reaumer's* thermometer, the zero marks the freezing point of water ; the boiling point is called 80.

Degrees of temperature below the zero points are generally indicated by the minus sign (—) placed before the number. Thus, — 40°, means a temperature 40° below zero.

8. *Other forms.*—Mercury freezes at about — 39° : temperatures below this point are measured by thermometers containing *alcohol.* Mercury boils at 660° : temperatures above this point are measured by the expansion of *solid bodies.*

When it is necessary to show very delicate changes of temperature, the *air thermometer* is used. This instrument has a variety of forms, but it consists essentially of a glass tube, terminating at one end in a bulb, the other end being open and inserted into a cistern of colored liquid. (See Fig. 119.) The liquid fills a part of the tube; the rest of the tube and the bulb above is filled with air. A graduated scale is placed behind the tube. The air expands or contracts with every change of temperature, and accordingly drives the colored liquid down, or allows it to rise in the tube. The motion of the liquid shows the change in the temperature.

Fig. 119.

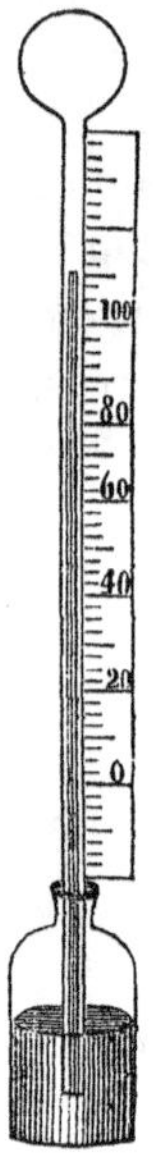

(84.) Temperature indicates the rapidity of vibration of the molecules of bodies: expansion indicates a change in their relative positions. The heat which produces the first is called sensible heat: that which produces the second is called latent heat. The sum of both quantities in any body, compared to the sum of both in some other body taken as a standard, is called the specific heat.

1. *Rapidity of motion and change of position.*—He who has a clear idea of the molecules [see (4.) 1], can distinctly imagine the multitude of these little bodies of which any larger body is made up, separated from each other by minute distances and in rapid motion. Now, heat can make them vibrate faster: it may also push them farther apart, or otherwise change their position; it can do nothing more. Then, when heat is being applied to a bar of iron, let the mind picture to itself these two effects; the molecules of the bar vibrating more and more swiftly, and at the same time being

pushed farther and farther apart. The first of these effects is manifested as temperature, the second as expansion.

2. *Sensible and latent heat.*—The heat which is expended in raising temperature is called *sensible* heat; it can affect the sense of touch. That which is used to produce expansion, is called *latent* heat; it does not affect the sense of touch. Now, the heat that goes into, or acts upon, any body, is divided into these two portions; one part sensible, the other latent.

3. *Specific heat.*—But different substances do not divide it alike; that is, if the same amount be added to two substances, one of them will devote more of it to temperature and less to expansion, than the other.

Let equal weights of water and mercury be placed over the same source of heat. The water divides the heat it receives into two parts, one to raise its temperature, the other to expand it. The mercury, receiving the same amount, divides it into two parts devoted to the same purposes, but the heat devoted to temperature is more than in water, while that devoted to expansion is less. We find that the temperature of mercury rises much faster than that of water: it takes thirty times as long to raise the water to a given temperature as it does the mercury. If it take thirty times as long, and one receives heat as fast as the other, there must be thirty times as much heat in the water as in the mercury when that temperature is reached by both. We see, then, that at the same temperature different substances may have very different quantities of heat in them. The relative quantities of heat in different bodies at the same temperature is called *specific heats.*

Water is the standard of specific heat. At a given

temperature it contains more heat than any other known substance. Its specific heat being 1, the specific heats of all other substances are fractional. The specific heat of mercury is .03. By this is meant, that when equal weights of mercury and water are at the same temperature, the mercury will contain only .03 as much heat as the water.

(85.) The expansion of a solid body will continue nearly uniform until its temperature has reached the melting point. The temperature then stops rising, while the expansion increases and continues until the solid is melted.

1. *The melting point.*—The temperature at which a solid body begins to melt, is called its *melting point.* At this point, the repulsive force of heat nearly balances the cohesion of the molecules, and enables them to move freely among themselves. The body becomes a liquid, and the change is called *liquefaction.* The melting point for different substances is not the same. Ice melts at 32° F.; mercury at 39°; iron at about 3,000° and platinum at about 5,000°.

2. *The temperature stops rising.*—If heat be applied to a vessel of ice at 32°, the ice will melt and the water formed will have the same temperature, 32°. So, too, when wax, or iron, or lead is melted, the liquid will have the same temperature as the solid which is melting.

3. *But the expansion increases.*—But in the case of all the substances above named, except ice, the expansion is greater at the melting point than before it was reached. The liquid fills more space than the solid from which it was formed. It should be so,

because the heat force is *all* expended to change the position of the molecules, whereas, before, a part of it was used up to produce a rise of temperature.

Ice contracts when melting; the water formed fills less space than the ice. In this case, likewise, all the heat applied is expended in *changing* the *position* of the molecules, but not in pushing them farther apart, for they occupy less space than before the change occurred. The change consists in throwing the molecules out of their crystalline arrangement. The water will continue to contract until it reaches a temperature of 39°, after which it expands.

Those who have attempted to melt ice or snow, for domestic purposes, remember how slow the process is. The amount of heat required to simply melt the snow without making it any warmer, is very great; the same amount applied to the water formed, would raise its temperature 142°. Hence the latent heat of water is said to be 142°.

(86.) The expansion of a liquid will continue gradual until the boiling point is reached, a temperature depending upon the purity of the liquid, upon the nature of the vessel in which it is heated, and upon the pressure it sustains. At the boiling point, the temperature stops rising, while the expansion greatly increases, and continues until the liquid is vaporized.

1. *The boiling point.*—The temperature at which a liquid begins to boil, is called its *boiling point.* At this temperature, the repulsive force of heat entirely overcomes the cohesion of the molecules, and drives them just as far apart as possible. The body rapidly becomes a vapor, and the change is called vaporization.

Liquids do indeed change to vapor at all temperatures. Even from freezing water, more or less vapor is ever slowly rising. This slow change is called evaporation. The boiling point for different liquids is not the same. Water boils at 212°, Alcohol at 173°, Ether at 95°. The boiling point for the same liquid is not the same; it depends upon three circumstances.

2. *It depends on the purity of the liquid.*—It is affected, first, by the presence of impurities in the liquid. The presence of some impurities raises the boiling point; of others, lowers it. Salt water, for example, boils at a higher temperature than pure water, while that which contains air, boils at a much lower temperature than that which contains none.

3. *It depends on the nature of the vessel.*—In an iron vessel, water will boil at a lower temperature than in one of glass. It is so because there is a stronger adhesion between water and glass, than between water and iron. The stronger adhesion requires a stronger heat to overcome it.

4. *It depends on the pressure.*—But the most important circumstance on which the boiling point of a liquid depends, is the pressure it sustains. This pressure is due to the atmosphere, to the weight of the liquid itself, and to any force which may be brought to bear upon it by artificial means. Whatever may be its cause, the effect of pressure is to raise the boiling point. It is well known that water boils at a lower temperature on the top of a mountain than at its base. It does so because the pressure of the air upon it is less.

This very important principle may be easily illustrated by experiment. For this purpose take a glass flask, or better, a bolt head (Fig. 120), and put into it water enough to fill the stem and a small part of the bulb. Invert it so that the water may be boiled by holding the bulb over the flame of a lamp. Boil it until the steam issues freely from the stem, and then, removing it from the flame, cork the stem at the same instant. The air has been driven out from the instrument, and nothing remains, to press upon the water, but steam. Turn the bulb upward so that the water may fill the stem; pour *cold* water upon the bulb, and the water inside will boil violently. Even when the tube has become so cold that it may be handled without inconvenience a fresh bath of cold water will cause the boiling to continue. It boils at the low temperature because the cold water, condensing the steam, removes the pressure from its surface.

Fig. 120.

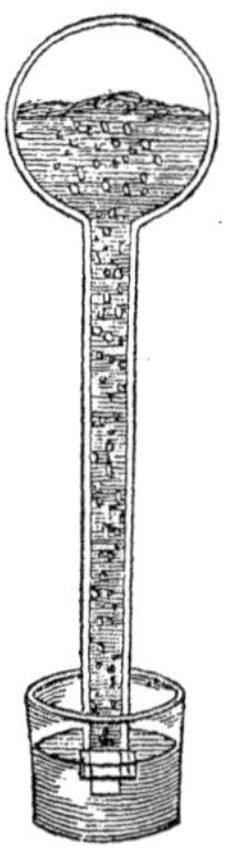

5. *The temperature stops rising.*—No matter how much heat may be applied to boiling water, its temperature can not be raised. Moreover, the temperature of the steam is always that of the water from which it is made.

By the following experiment these facts may be illustrated. Water is placed in an open vessel V (Fig. 121). Into the water is plunged the bulb of an air thermometer T, whose tube is bent twice at right angles for the purpose. While the water is being heated by a lamp-flame, the gradual sinking of the fluid in the stem of the thermometer shows the increase of temperature; but when the water fairly boils, the fluid stops sinking, showing that the temperature no longer rises.

The fluid will remain motionless until the water in the vessel has been changed to steam. Let the bulb be lifted into the steam above the water; no change occurs in the height of the fluid in the stem, hence the temperature of the steam must be the same as that of the water.

Fig. 121.

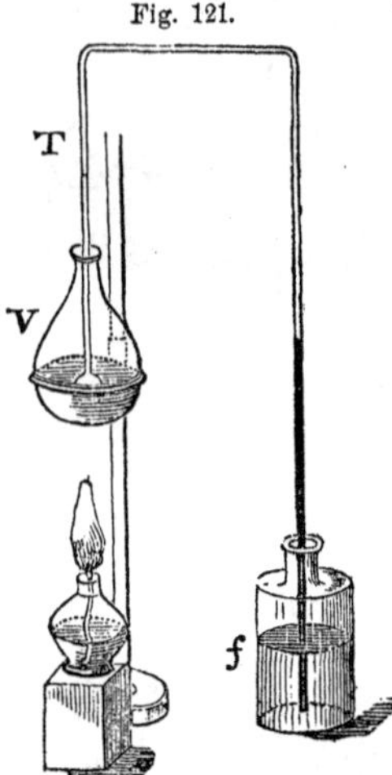

6. *But the expansion increases.*—If all the heat applied to a boiling liquid is expended to produce expansion, we may expect that this effect will be more rapid than when a part of it was used to raise the temperature. This inference is abundantly verified. Steam fills about 1700 times as much space as the water from which it was formed.

The amount of heat required to expand water into steam, in other words, the latent heat of steam, is very great. By accurate experiment it has been found to be 972° F. The steam and the water are of the same temperature, yet there is an excess of heat-force in an ounce of steam which, if applied to one ounce of water, would be sufficient, were it possible, to raise its temperature to 972; it will raise the temperature of *nine ounces* of water 108.

(87.) The heat-force which has been required to produce expansion, will be reproduced when the expanded body again contracts.

1. *Heat restored after being used.*—The following experiment very satisfactorily shows that heat is re-

quired to produce expansion, and that it is again given off when contraction occurs. The bulb of an air thermometer is placed in a receiver over the plate of an air-pump; the stem, passing through the top of the receiver, is bent twice at right angles, and is filled with its colored fluid to a height very carefully marked. By a few rapid strokes of the piston, the air is partly exhausted from the receiver; that which remains *expands*, and the rising of the fluid in the thermometer shows that the bulb is at the same time *cooled*. If, now, air be allowed to return to the receiver, the air inside becomes *more dense*, and the sinking of the fluid in the thermometer shows that heat is again given off.

Numerous familiar facts illustrate this principle. If water evaporates from the hand, it cools the hand, because the hand furnishes the heat to expand the water into steam; and when vapor condenses upon the hand, the hand is warmed, because the vapor gives to it all the heat which had been used to keep the water in the form of steam.

Bodies in contact with melting snow or ice, are cooled, because they must furnish heat to change the solid to the liquid form. But bodies near to freezing water are warmed, because the water then gives up the heat which had been needed to keep it in a fluid form.

§ 4. ON THE STEAM-ENGINE.

(88.) The elastic force of steam is applied to mechanical purposes by means of a steam-engine. The essential parts of this machine are, 1st, the boiler in which steam is generated; 2d, the cylinder in which the steam is made to move a piston; 3d, the crank by which the piston turns a wheel. Engines are either high-pressure or low-pressure.

1. *The elastic force of steam.*—When steam is formed at a temperature of 212°, its elastic force is just equal to the pressure of the atmosphere, or 15 lbs. to the square inch. If taken out into another vessel, preserving its temperature and density, it would exert a pressure of 15 lbs. to the inch. By subjecting water to a greater pressure, its boiling point is raised, and the elastic force of the steam will be increased. The Marcet's globe illustrates this principle. It consists of a metallic globe (Fig. 122), which is furnished with a long glass tube and scale T, a stop-cock S, and a thermometer A, whose bulb is inside the globe. In the bottom of the globe is a little mercury, into which the end of the tube T dips, and above the mercury is a quantity of water. The water is boiled until the air is driven out of the open stop-cock. At this moment, the elastic force of the steam is just 15 lbs. to the inch. The stop-cock is now closed: the thermometer at once shows a rise of temperature, and at the same time the mercury begins to rise in the tube, showing an increase in the force of the steam. When the temperature of the boiling water has reached 249.5°, the expansive force of the steam is equal to two atmospheres, or 30 lbs. to the inch, and at 306° it is five atmospheres, or 75 lbs. to the inch.

Fig. 122.

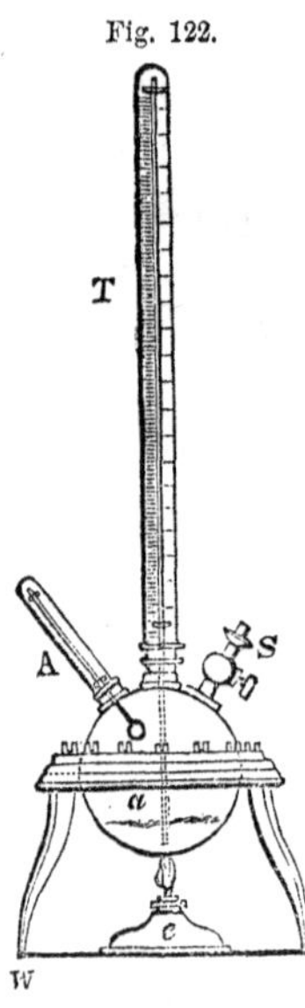

If, now, this elastic force of steam can be made to act alternately upon opposite sides of a piston, it will knock it back and forth, from one end of a cylinder to the other with power enough to move any amount of other machinery. This is accomplished in the *steam-engine.*

2. *The boiler.*—The boiler of a steam-engine is usually made of plates of wrought-iron riveted together in the form of a cylinder. In the best forms, there are tubes which run lengthwise through the body of the boiler, through which the flame and hot gases from the fire may pass. The water in the boiler surrounds these tubes, and is rapidly heated by them. The steam thus formed in the boiler collects above the water, and by its pressure raises the boiling point until, when its elastic force is sufficiently great, the steam is allowed to pass through a pipe to the cylinder.

Fig. 123.

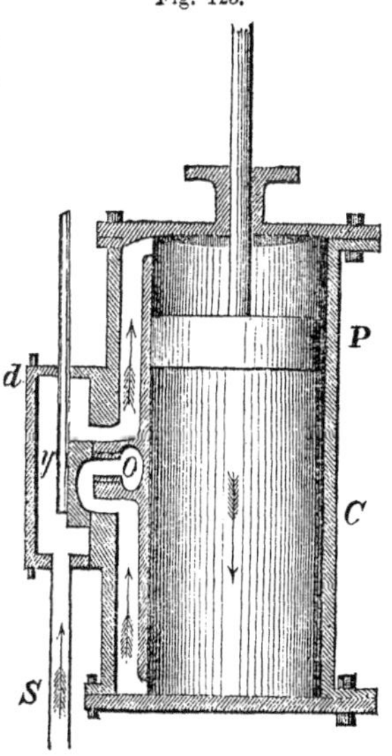

3. *The cylinder.*—The arrangement of the cylinder and piston are shown in Fig. 123. The pipe which brings the steam from the boiler enters a box d, from which two tubes lead, one to the top, the other to the bottom of the metallic cylinder C, in which the piston P, moves. Another tube leads from this box out into the air, or away to another vessel, where the steam, after having moved the piston,

may be condensed. A sliding valve, Y, is so arranged in the box as to always close one of the pipes leading to the cylinder and leave the other open. If the upper tube is open, as represented in the figure, the steam will enter above the piston and push it to the bottom of the cylinder; if the lower tube is open, the steam will enter below the piston and push it to the top. In either case the steam on the opposite side of the piston will be pushed out of the cylinder, through the other tube and the pipe, O, leading from the cavity under the sliding valve. When the steam, entering through the lower tube, has pushed the piston to the top of the cylinder, the valve is pushed down to cover the end of that tube, leaving the end of the other uncovered, so that the steam may pass through it to act above the piston. By this means the piston will be alternately pushed back and forth from one end of the cylinder to the other.

4. *The crank.*—By this simple motion, back and forth, the piston turns a wheel by means of a crank. To the piston-rod, A (Fig. 123), another rod is joined by a hinge; the other end of this rod turns a wheel, from which motion may be communicated to others by bands or cogs.

Besides these three important parts of the steam engine, there are numerous other appendages for particular purposes, such as a safety-valve attached to the boiler to regulate the pressure of steam in it; the governor to regulate the supply of steam to the cylinder; the fly-wheel, a heavy wheel whose inertia causes the motion of the machinery to be steady. (See Silliman's Physics.)

5. *High and low pressure engines.*—The different forms of steam-engines are almost as numerous as the

machinists who make them, or as the variety of purposes to which they are applied. There are, however, two general classes, the *high-pressure* and the *low-pressure* engines.

In the high-pressure engines the steam, after moving the piston, is thrown out from the cylinder into the air. In the low-pressure engines, the steam, after moving the piston, is taken off to a vessel called the condenser, in which it is changed into water. The first is called *high* pressure, because the steam which moves the piston must push the steam from before the piston out into the air, which presses it back with a force of 15 lbs. to the inch. To do this evidntly requires a pressure of 15 lbs. to the inch *higher* than in the other class, in which the steam escapes into a vacuum, and of course exerts no pressure against the piston.

CHAPTER VIII.

ON ELECTRICITY.

INTRODUCTION.—APPLICATION OF THE FUNDAMENTAL IDEAS.

(89.) READ (4). A constant and opposite action of attraction and repulsion among the molecules of bodies gives rise to the phenomena of electricity.

Of the nature of electricity very little is definitely known; but since the kindred phenomena of light and heat have been found to be the result of vibratory motions among the molecules of bodies, the tendency is to regard electricity also as the effect of molecular vibrations of some kind. But whether we adopt this view, or still cling to the old theory, which regards electricity as a weightless fluid in the pores of all bodies, we may describe it truly as a manifestation of *attraction* and *repulsion* acting upon the molecules of a body, thus producing an effect upon the body itself.

§ 1. OF FRICTIONAL ELECTRICITY.

(90.) Electricity may be produced by friction. The electrical machine is an apparatus for this purpose. It may be detected by instruments called electroscopes, showing its action as two opposite forces—attraction and repulsion. Its intensity may be measured by instruments called electrometers. It is governed by two laws:—

1st. Electricities of the same kind repel each other; of different kinds attract.

2d. The force of the attraction or repulsion is inversely as the square of the distance between them.

1. *Electricity produced by friction.*—If a well-dried glass tube—a lamp-chimney for example—be thoroughly rubbed with a flannel cloth, it will be found to have new and curious properties. Hold it near the face, and a feeling will be experienced as if a gentle breeze were blowing against the cheek: bring it nearer, and perhaps a prickling sensation will be felt, and it may be that a crackling sound will, at the same time, be heard; or approach it toward some very light substances, such as delicate bits of loose cotton and they will rush toward it, and remain for a little time clinging to it. These various effects show the presence of electricity: the friction of the flannel upon the glass has produced it.

2. *The electrical machine.*—The electrical machine is an apparatus for producing electricity by friction. It is represented in Fig. 124. Its principal parts are, 1st, a body upon whose surface electricity is to be evolved; 2d, the rubber by the friction of which electricity is produced; and 3d, the conductor on which the electricity may be accumulated. In the form shown by the figure, the first of these parts consists of a thick glass plate P, to be turned by a crank. The rubber R is made of leather, covered with an amalgam made of mercury, tin, and zinc. Two such pieces of leather are pressed, one against each side of the plate, by means of a brass clamp, which is supported

upon a glass pillar. The conductor, or as usually called the *prime conductor*, C, is a brass ball, or a cylinder with rounded ends, mounted on a glass support. Con-

Fig. 124.

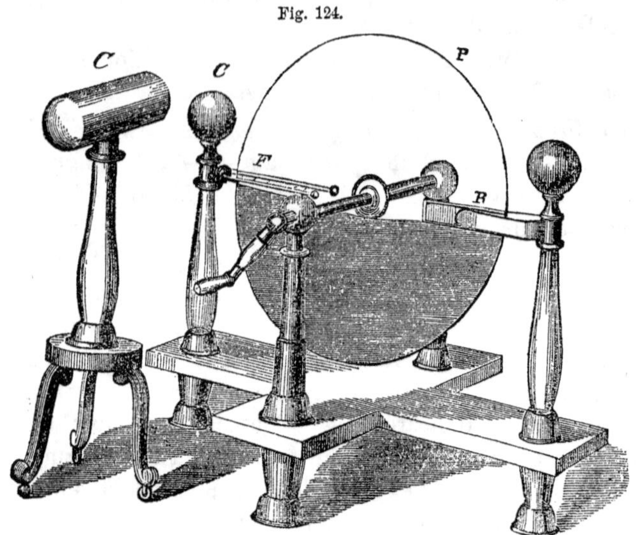

nected with the prime conductor, is a brass fork F, one prong of which is on each side of the plate, with many sharp projecting points reaching toward the glass.

By turning the crank, the friction of the rubber upon the plate evolves electricity, which remains upon the surface of the glass until it is brought around to the fork: it is there taken by the points, and it passes over the prongs to spread itself over the surface of the prime conductor. The glass support prevents it from leaving the conductor. When the machine is in operation the rubber is connected with the floor by a chain. All parts of the machine must be free from dust and thoroughly dry.

When a machine of this kind, of medium size, is in successful operation, the effects of the glass tube are experienced in a far greater degree. The face or the back of the hand will feel the breezy or prickling sensation at a distance of several inches from the conductor: all light bodies held near it, immediately fly to its surface, and if the knuckle or a brass ball be brought near, bright and zigzag sparks may be drawn through a distance of from one to two inches, the light being accompanied by a sharp report.

3. *Electricity detected by electroscopes.*—When the force of the electricity is slight, there should be some convenient way of showing its presence. Any instrument for this purpose is called an *electroscope.* The simplest form is called the *pith-ball* electroscope. It consists (see Fig. 125) of a ball of pith from the cornstalk, or elder, hung by a slender silk thread from a glass support. This little ball will instantly announce the presence of electricity by moving toward the body which contains it.

Fig. 125.

4. *Two opposite forces.*—Electricity shows its presence both by attraction and repulsion, for if the pith-ball of the electroscope be brought near to the prime conductor of the electrical machine, it will fly toward it, but on coming in contact with it, will as instantly leap away again.

Now, rub a glass rod with flannel, and hold it near the pith-ball which has been repelled by the conductor; the glass rod will also repel it; but if a stick of sealing-wax be used in place of the glass tube, the pith-ball will be strongly attracted. Notice, that the pith-ball is *repelled* by the electricity of *glass*, and *attracted* by the electricity of *sealing-wax*. It is thus seen that the electricities of glass and sealing-wax are not alike. To distinguish them from each other, that which is produced on glass by the friction of flannel is called *positive* electricity; that produced upon sealing-wax is called *negative* electricity.

It is found to be impossible to develop one of these forces without the other also. The positive force always appears on one of the bodies rubbed together, and the negative upon the other.

5. *The forces measured by electrometers.*—The simplest form of the electrometer is represented in Fig. 126. It consists of a brass standard, with a graduated semicircle, on the center of which moves an index of very light wood, carrying a pith-ball at its lower end. When not in use, the pith-ball hangs in contact with the standard, but when the standard is brought near to an electrified body, the pith-ball is instantly repelled. The arc through which it moves is taken as the measure of the force. (See Silliman's Phys., p. 538.)

Fig. 126.

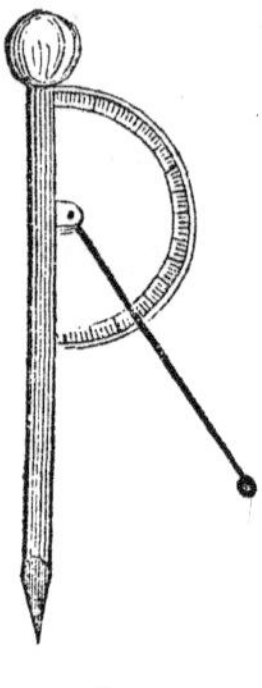

6. *The first law.*—We have seen that positive electricity is produced by friction on glass, and that the opposite force is evolved by friction on sealing-wax. Now, let two pith-balls be suspended by silk threads so as to be in contact. Thoroughly rub the glass tube; bring it in contact with the balls; they both receive positive electricity from the tube, and it will found that they will no longer remain in contact. We learn from this experiment that two bodies with the *same kind of electricity repel each other*.

Again: let the sealing-wax be thoroughly rubbed and brought near to the two pith-balls while they are repelling each other, and they will both fly toward it. We learn from this experiment that bodies with *different kinds of electricity attract each other.*

This law furnishes an easy test by which to find out which kind of electric force is, in any case, produced. Is the prime conducter of the electrical machine positive or negative? To decide this question, rub the glass tube; bring it in contact with the pith-ball of the electroscope; the electricity of the ball is thus known to be positive. Now, bring it near the prime conductor of the machine in operation; it is *repelled.* The electricity of the conductor is positive. The electricity of the rubber is negative, because, if the chain be removed, and the electrified pith-ball be brought near the brass mounting of the rubber post, it will be attracted.

7. *The second law.*—If the standard of the electrometer be brought in contact with an electrified body, the index will be thrown along the graduated arc to a greater or less distance, as the electric force is stronger or weaker. By carefully conducted experiments, these distances may be compared, and it is found that the strength or intensity of the force is inversally as the square of the distance. The attraction of the prime conductor for a body at a distance of two inches is only $\frac{1}{4}$ as strong as at a distance of one inch.

(91.) A charged or electrified body, acting through a non-conductor upon an insulated conductor, polarizes it. This action is called induction.

1. *A charged body.*—Whenever by friction, electricity

is developed upon the surface of a body, the body is said to be electrified, and if, by bringing another body in contact with it, electricity is imparted, the body which receives it is said to be charged. Thus, the glass tube, when rubbed, becomes *electrified;* the pith-ball of the electroscope, coming in contact with the glass, takes electricity from it and becomes *charged.*

2. *A non-conductor.*—Some bodies allow electricity to pass freely over their surfaces; such bodies are called conductors: others will not allow electricity to pass freely over them; these are called non-conductors. If a brass rod be held in contact with the prime conductor of a machine, it will be found impossible to charge it; a glass rod held in the same way, will not prevent the charge from accumulating. The brass allows the electricity to pass into the person; the glass does not: brass is a conductor; glass is a non-conductor. The metals, as a class, are good conductors. Beside glass, we notice silk, india rubber, and dry air, as being among the best non-conductors.

3. *An insulated body.*—Whenever a body is quite surrounded by non-conductors, it is said to be insulated. The conductor of the machine is insulated by resting upon a glass support. A body which is not insulated can not be charged.

4. *A charged body polarizes an insulated conductor.*—A body is said to be polarized when the two opposite electricities both exist upon its surface. To illustrate this important condition, let an insulated metallic ball be connected with the prime conductor of the electrical machine, and let a small insulated conductor be placed near it (see Fig. 127). When the ball is charged, the

motion of the pith-balls fastened to the small conductor, shows that it is also charged, and if its electricity be tested, it will be found to be positive at one end and negative at the other. Both electricities are developed upon its surface at the same time, and the body is said to be polarized. The action of the ball, by which this body is polarized, is called *induction*.

Fig. 127.

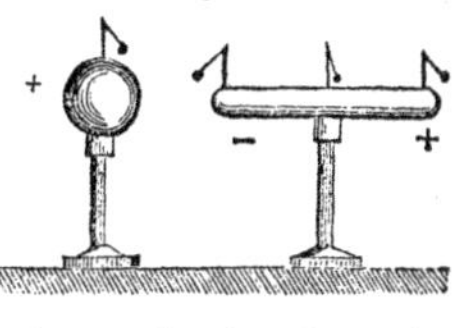

If we examine the condition of the polarized body more carefully, we find that in the end next to the ball there is *negative* electricity, and in the distant end there is *positive* electricity. This is always true: when a body is electrified by induction, the end or side nearest the charged body is always in a condition *opposite* to that which develops it.

When the insulated conductor is near to the ball, the induction is strong: the greater the distance between them, the weaker it becomes, until, at a certain distance, it can no longer be detected.

If, when the conductor is polarized, one end be touched with the finger, the entire surface remains charged with the opposite electricity. It will remain charged even, when taken beyond the influence of the body which polarized it.

(92.) A series of insulated conductors, placed end to end, near each other, may be all polarized by bringing a charged body near to one of them. Faraday's theory explains induction by supposing the molecules of a body to be polarized from each other in the same way.

1. *A series of conductors polarized.*—Let a number of small insulated conductors be placed end to end,

near together, with one end of the first one near to a brass ball connected with the prime conductor of the machine (see Fig. 128). The motion of the pith-balls will show that they are all polarized. The effect will be greater if another brass ball, connected with the rubber of the machine, is placed at the other end of the series. The positive and negative electricities are on opposite ends of each conductor. All the ends toward the positive ball are negative; all the ends in the other direction are positive.

Fig. 128.

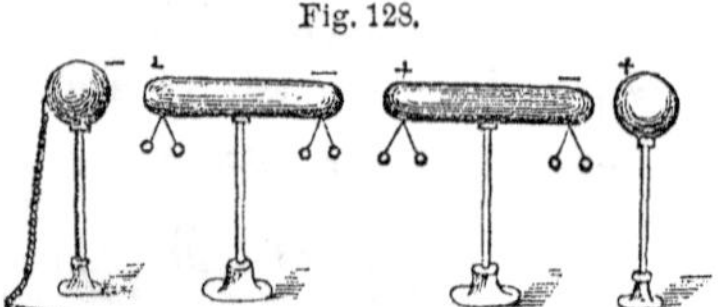

2. *The Theory of Induction.*—Now the molecules of one of these conductors are as truly separate bodies as the conductors themselves, and as one electrified conductor may polarize another, so one of these molecules, acting through the minute distance between them, may polarize another. This polarizing influence passes from one molecule to another, until all the molecules of the body are thrown into this condition, each molecule having opposite electricities on its opposite sides.

The theory goes further, and supposes that the molecules of conductors discharge their forces easily into one another, while those of non-conductors do not. For this reason, the molecules of the air between the charged ball and the end of the conductor are polarized and *retain* their electricities, while the molecules of the conductors, as fast as they are polarized, give their electric forces to their neighbors. The positive force given from one to another, in one direction, accumulates at one end of the conductor; the negative force, given from one to another in the other direction, accumulates at the other end.

(93.) The Leyden-jar is an apparatus for accumulating electricity by induction. It may be charged by bringing one of its coatings in contact with a charged body, the other being in contact with conductors. It may be discharged by making a conducting communication between its two coatings. The Leyden battery consists of several Leyden-jars connected.

1. *The Leyden-jar.*—The Leyden-jar (Fig. 129) consists of a glass jar, coated both inside and outside with tin-foil, to within a few inches of the top, and provided with a cover of hard dry wood, through which passes a brass rod, with a ball upon its upper end, and a chain reaching from its lower end to the bottom of the jar.

Fig. 129.

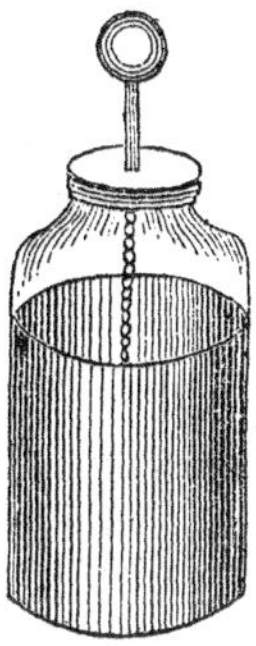

It will be seen by this description, that in this instrument there are *two conducting surfaces*, separated from each other by a *non-conductor*.

This idea may be embodied in a variety of forms, any one of which will act on the principle of the Leyden-jar. Thus a pane of glass, coated with tin-foil on both sides, to within a little distance of the edge all around, has the essential parts of the Leyden-jar. A glass goblet partly full of water and grasped by the hand, illustrates the same idea: the glass, a non-conductor, separates two conducting surfaces—the water on the inside, and the hand upon the outside.

2. *It may be charged.*—By bringing the ball of the Leyden-jar in contact with the prime conductor of the

machine, positive electricity passes into the inside coating. This positive electricity polarizes the outside coating, causing its surface next the glass to be negative, and the other to be positive. If in contact with a conductor, this positive electricity will pass off, and thus leave the outside coating permanently charged with negative electricity. When by this action the two coatings have opposite electricities, the jar is said to be *charged.* It may be removed from the prime conductor and remain charged, because the two forces hold each other by acting through the glass. The jar may be handled without danger, if care be taken not to touch the ball and the outside at the same time.

The jar is charged with positive electricity when that force is upon the inside; it is charged with negative electricity when negative force is upon the inside.

3. *It may be discharged.*—When a conducting communication is made between the two coatings of the jar, the two opposite forces come together, neutralize each other, and the jar is said to be discharged. The conducting communication may be made in many ways. The *discharger* is a convenient instrument for the purpose. It consists of two bent brass arms, with a ball upon one end of each, the other ends being fastened by a joint to a glass handle. Taking hold of the glass handle, bring one ball in contact with the outside of the jar, and the other near to the knob; a bright spark and a sudden report announce the discharge.

The coated glass plate and the goblet of water, mentioned before, may be charged and discharged in the same way as a Leyden jar. To charge the goblet, for example, let a chain from the prime conductor of the machine hang into the water; grasp the outside of the glass while the machine is in operation. Positive electricity will be given to the water; negative electricity will be induced upon the hand, and the goblet is thus charged. Now with the other hand try to remove the chain: the moment the chain is touched, a slight *shock* will be felt, announcing the discharge which occurs.

4. *The Leyden battery.*—The larger the surface of the coatings of the jar, the more powerful will be the charge accumulated. We can obtain a larger surface by using a larger jar, or it may be done by taking several small ones and joining their surfaces by conductors. In the last case, the Leyden battery will be formed. When the inside surfaces are all connected by conductors reaching from knob to knob, and the outsides all joined by standing the jars on a metallic surface, the battery may be charged and discharged as a single jar. It is equivalent to a single jar large enough to have the same extent of surface.

(94.) The electricity of the atmosphere is of the same nature as that produced by friction. Lightning is the discharge of oppositely charged clouds, illustrating, on a grand scale, the action of a Leyden-jar.

The aurora is doubtless produced by electric discharges taking place in the rarefied air of the upper portions of the atmosphere.

1. *Electricity of the atmosphere.*—The atmosphere is very generally in an electrified condition. This may be shown by raising a metallic rod to a considerable height above the ground, having an electroscope fastened to its lower end, which should be insulated. A *sensitive* electroscope will usually indicate positive electricity, its intensity increasing as the air from which it

is drawn is higher. In its ordinary state, the electricity of the atmosphere is always positive: stronger in winter than in summer, and during the day than the night. In cloudy weather the electrical state is uncertain, sometimes changing from positive to negative and back again in a few minutes. On the approach of a thunder-storm these changes follow each other, at times, with remarkable swiftness.

2. *It is of the same nature as frictional electricity.* —The bright flash and loud report which announce the discharge of a Leyden-jar or battery, can not have failed to remind one who has observed them, of the brighter flash and louder report of atmospheric lightning and thunder. These grand and sometimes awful displays of electricity, are caused by the same agent which, produced on a glass tube, lightly pricks the cheek or attracts a pith-ball.

To Dr. Franklin belongs the immortal honor of proving the identity of electricity and lightning. A kite was the simple instrument employed by this man of genius. Having made a kite by stretching a silk handkerchief over two sticks in the form of a cross, he went out into a field, accompanied only by his son: raised his kite; fastened a key to the lower end of its hempen string; insulated it by fastening it to a post by means of a silk cord, and anxiously awaited the approaching storm. A dense cloud, apparently charged with lightning, soon passed over the spot where he stood, without causing his apparatus to give any sign of electricity. He was about to give up in despair, when he caught sight of some loose fibers of the hempen cord, bristling up as if repelled. He immediately presented his knuckle to the key, and received an electric spark. The string of his kite soon became wet with the falling rain; it was then a better conductor, and he was able to obtain an abundance of sparks from the key. By this experiment he furnished a decisive proof of the identity of lightning and electricity.

3. *Lightning is the discharge of oppositely charged clouds.*—Clouds are often charged with electricity. When two of them, with opposite kinds of electricity, come near enough together, they will act like the two charged coatings of the Leyden-jar, the air between them being a non-conductor like the glass. When the charge rises high enough, a discharge takes place: the spark of the discharge being a flash of lightning, and its report a thunder peal. Considering the large extent of cloud surfaces discharged, we need not be surprised at the magnitude of the spark, nor at the deep intensity of the sound.

When the discharge is not hidden by clouds, we can trace the whole length of the spark, and we witness *chain-lightning;* but at other times the spark is behind the clouds; we see only the light of the discharge spread over the surface of the clouds, and this gives rise to what is called *sheet-lightning.*

At times the earth and a cloud are the two charged surfaces, and a discharge takes place between them. Such discharges are the source of danger to life and property. Animals, trees, buildings, all these are better conductors than air, and electricity always chooses the best conductors in its passage. In going from a cloud to the earth it takes these bodies in its way; animals are often killed, trees shattered, and buildings torn to pieces or set on fire.

4. *The aurora.*—This curiously beautiful phenomenon

consists of a diffuse light, somewhat like the morning or the evening twilight, seen in the northern sky. It exhibits a great variety of appearances. Sometimes it looks much like the dawn of morning seen in the north instead of the east. Sometimes it takes the form of an arch, like a rainbow, but without its colors. At other times slender columns of delicate light, pointing upward from the northern horizon, not always stationary, but often, on the contrary, leaping up and down with swift and varied motions, as if engaged in a merry dance.

In the southern hemisphere an aurora is also seen in the southern horizon. To distinguish these two auroras, that in the north has been called the *aurora borealis*, while that in the south is the *aurora australis*.

5. *It is produced by electric discharges in rarefied air.*—There is still much uncertainty about the cause of the aurora, but late investigations leave no doubt as to its electrical nature. From all the facts gathered, it seems to consist of beams or discharges of electricity, between the earth and the upper regions of the atmosphere.

When electricity discharges through air of the usual density, it takes the form of a spark, the light being intense and nearly white. If passed through a glass vessel, in which the air is rarefied, the light is more diffuse and tinged with a rosy hue. If the air be still further rarefied, the light becomes very diffuse, spreads readily through a great distance, and its color becomes a deep rose or purple.

Now the air of the upper atmosphere is much rarefied, and we should infer that electric discharges there would give a diffuse light of various colors. Such is the observed character of the aurora. (See Smithsonian Report, 1865, p. 208.)

(95.) A body having points projecting from its surface can not be charged even when insulated. Or if a pointed conductor be held toward its surface it will prevent a charge from accumulating, by drawing the force away silently. Upon this principle, buildings are protected from the effects of lightning by lightning-rods.

1. *The effect of points.*—It is found to be impossible to charge a conductor when there are sharp points on its surface, or held near to it. To illustrate this curious effect of points, fasten a pointed wire to the prime conductor of the electrical machine, and the sparks, which before could be drawn from it in abundance, cease altogether, and even pith-balls fail to detect the presence of the force. Or take the pointed wire in the hand and present its point to the prime conductor, within a few inches of its surface; not a spark can be drawn from it, nor will the pith-balls show either attraction or repulsion. The discharge is silently effected by the air in front of the points. Its molecules become polarized, and are first attracted to the point and then repelled. On coming in contact with the point, they take electricity from it and move away: others being polarized are attracted, receive electricity, and pass away. Thus the electricity of the body is silently carried off from the point. That such currents of air do really exist, may be proved by various experiments. If, for example, the cheek, or the back of the hand, be held near to the point, the breeze will be felt: or if the small flame of a lighted taper be held just in front of

the point on the prime conductor, it will be blown away from it, and may even be extinguished.

The discharge takes place from the point because the charge being more intense there than elsewhere, the polarization of the air is greater there than at any other part of the body.

2. *Lightning-rods.*—We have seen that because buildings are better conductors of electricity than air, they are liable to injury from strokes of lightning, when the discharge takes place between the cloud and the earth. But since pointed conductors silently discharge the force from a charged body, why not disarm the cloud of its lightning by the use of pointed metallic rods? This question was no sooner suggested to the practical mind of Franklin, than a trial was made, which verified his bold conjecture.

Conductors for the purpose of protecting buildings from the effect of lightning, are called lightning-rods. They should be made of metallic rods, pointed at the upper end, reaching several feet above the highest part of the building which they are designed to protect, and downward, without interruption, into the ground below its foundation.

(96.) The effects of frictional electricity are mechanical, chemical, and physiological.

1. *Mechanical effects.*—We have already had abundant illustrations of *motions* caused by the electric forces. Poor conductors are also pierced or torn by the electric discharge. To illustrate this by experiment, let the charge of a Leyden-jar be passed through a piece of card-board; the card will be pierced with a burred or ragged perforation. This effect is produced on a

large scale by the lightning stroke; even rocks are sometimes shattered, while trees are often splintered from top to root, and their fragments scattered far and near in all directions.

2. *Chemical effects.*—The chemical effects of electricity are shown through the agency of the heat which it develops. To illustrate by experiment: wrap the ball of a Leyden-jar with loose cotton, and sprinkle upon this, very finely powdered resin. This done, charge the jar powerfully, and then discharge it by bringing first one ball of the discharger in contact with the outside of the jar, and then the other a little above its hooded knob. The discharge takes place through the resin, and sets it on fire. Buildings are sometimes set on fire by the lightning-stroke.

3. *Physiological effects.*—The effect of electricity upon the human system is peculiar and startling. No description can give a correct idea of it: it must be experienced by one who would know what it is. Let a person place one hand upon the outside surface of a lightly charged Leyden-jar, and with the other hand touch its knob. He will find that his own will can no longer control his muscles: his hands are, on the instant, suddenly jerked, while a peculiar and almost indescribable sensation is felt in the wrists and arms.

Many persons by joining hands may form an unbroken connection between the two coatings of the jar, and at once experience these effects.

§ 2. OF MAGNETIC ELECTRICITY.

(97.) Magnets are either natural or artificial, and may be made in different forms; but in any form the magnetism is stronger at the ends than in the middle. The ends are called poles.

1. *Magnets.*—Bodies that attract iron in preference to other metals, are called magnets. They are usually made of steel. To illustrate their peculiar preference for iron, let some iron filings be mixed with some filings of brass; bring one end of the magnet among the filings, and on removing it, great numbers of the iron particles will be seen clinging to it, while the brass particles are all left behind.

2. *The natural magnet.*—Fragments of an ore of iron are sometimes found which have the properties of a magnet. Such a fragment is a *natural magnet* or *loadstone.*

3. *The artificial magnet.*—If a bar of iron or steel be rubbed against a magnet it will become magnetic; it will then be an artificial magnet. Whether it remains magnetic for any length of time depends upon its hardness. Soft iron or steel will lose its magnetic properties quickly; hardened iron or steel will retain them.

4. *They are made in different forms.*—The two most important forms of magnet are the *straight* or *bar* magnet, and the *horseshoe* magnet. These names are descriptive: the bar magnet is a straight bar of steel; the horseshoe magnet is a magnet whose shape is that of a horseshoe, or the letter U; its ends are thus brought near together.

5. *Their force is stronger at their ends.*—If a bar magnet be rolled in a bed of iron filings, large clusters of them will be found clinging to its ends, their numbers getting less toward the middle of the bar, where very few, if any, will be held. By this experiment we

learn that the magnetic forces are not equally distributed over the surfaces of magnets, but, on the contrary, that they are strong at the ends and weak or neutral in the middle. The ends are called poles, one being a *north* pole, the other a *south* pole.

(98.) Magnetism shows itself both by attraction and repulsion, obeying the following law: poles of like names repel each other; those of different names attract.

1. *Attraction and repulsion.*—Iron which is not magnetic will be attracted equally by both poles of a magnet; it is not so when two magnets act upon each other. By presenting one end of a magnet to the north pole of another, it will show an attraction for it, while the other end being presented to the same pole will repel it. Thus magnetism, like electricity by friction, shows itself by both attraction and repulsion.

2. *The law.*—When we examine the subject more closely, we find that magnetic attraction and repulsion is also governed by the same law as electricity. Thus, if we bring the *north* poles of two magnets near to each other, they repel each other: two *south* poles manifest the same effect. But if we bring a north pole near to a south pole, an attraction instantly springs up between them, and if allowed to touch each other they will cling together: one may even lift the other by the strength of the attraction. It is evident from these simple experiments that poles of the same name repel, while those of opposite names attract each other.

(99.) A magnet, like a charged body, will polarize a bar of iron brought near to one of its poles, always inducing magnetism of the opposite kind in the end next to it. The polarizing influence may extend through several bars placed end to end.

It is supposed that every molecule of a magnet is in a polarized state, the north polarity being on the same side of them all, and the south polarity on the other side.

I. *A magnet will polarize a bar of iron.*—If a bar of iron *n s* and a magnet N S (Fig. 130), be placed end to end, the iron itself becomes a magnet. That it is a magnet may be shown by its power to attract or repel the poles of another magnet. Both kinds of magnetism are developed in it, and hence we call it polarized.

Fig. 130.

N S n s

If now, we examine more carefully, we notice that one end of the bar is near the south pole of the magnet, and we find by experiment, that the opposite end is likewise a south pole. Hence, that end of the bar next to the magnet must be a north pole. Each pole of a magnet will always in this way induce the opposite kind of magnetism in that end of the bar nearest to it.

Unlike frictional electricity, there is no discharge of magnetism when opposite kinds are brought together: the polarization takes place even when the bar is in contact with the magnet, and if the bar be made of steel, the polarity remains after the magnet is removed.

2. *Several bars may be polarized.*—A second bar may be placed with one end near to the first, and be found to be polarized; so a third may be polarized by the second: the series may be continued further, but the force is less in each successive magnet. To illustrate

by experiment: from the north pole of a strong bar magnet, hang a key, from the lower end of this one a smaller key may be hung, a third still smaller will be held by this, and a tack will cling to the lower end of the last. The series of keys and nails has become a series of magnets, each with its north and south pole, their north poles all directed downward (see Fig. 131).

Fig. 131.

S

N

3. *All the molecules of a magnet are polarized.*—Now the molecules of a magnet are as truly separate from each other as the several magnets in the series just described. And it is thought that each molecule is a magnet, with its north and its south pole. Acting through the minute distances that separate them, each one is polarizing its neighbors, and hence, like the series of bars, their north poles must be all arranged in one direction, their south poles in the other. Both kinds of magnetism act upon each separate molecule, and keep it in a magnetic state. There is no transfer of the force from one molecule to another as there is of electricity in a charged body, so there can be no discharge of magnetism. A magnet, like a body charged with electricity, may *polarize* another, but it can not, like the charged body, become neutral by giving up its force.

Why then is the middle of a magnet neutral, while only toward its ends do the forces show themselves? Not because the force of one kind leaves the molecules of one end and goes to the other, but because each molecule, from one end to the other, is exerting this force in the *same direction.* If a row of boys stand close together, and all push in the same direction, those at the end of the line will receive the greatest pressure; so the molecules of a magnet near the end toward which either force is acting, will be endowed with the strongest magnetism. In the middle of the series, the two forces are equal and in opposite directions, and must neutralize each other.

(100.) If a bar magnet be supported so as to move freely in a horizontal direction, it will rest only when its poles point north and south or nearly so; its variation is subject to both annual and diurnal changes.

Fig. 132.

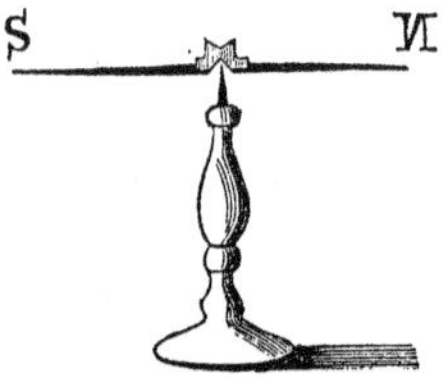

S N

1. *If a bar magnet be supported.*—A magnet may be supported in three ways so as to have free motion. It may be balanced upon a pivot. (See Fig. 132.) It may be hung from a fixed support by a fine thread tied about its middle point. (Fig. 133.) Or it may be, for purposes of simple experiment, fastened to a cork and laid upon water.

Fig. 133.

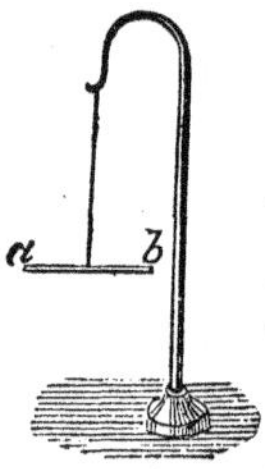

a b

2. *It will point north and south.*—The magnet supported in either of the ways mentioned, will swing back and forth, until it finally settles to rest, and it will then be found pointing north and south. The end which points north is called the north pole: it is evident, however, that its magnetism is of the kind opposite to that of the north part of the earth.

A slender bar magnet thus balanced is called a *magnetic needle.* Such a needle is used by mariners to direct them in their long voyages across the ocean. For this purpose, it is placed over a card upon which the "points of compass," north, south, east, west and others, are marked, and for protection, put into a box hung upon pivots, so that it will keep the needle in a horizontal position amid all the rolling or plunging motions of the ship. Such an arrangement is called the *mariner's compass.*

3. *Its variation.*—While it is true that the needle points in a direction which may be described as north and south, we must not understand that this description is exact. Indeed the needle seldom points *exactly* north and south. There are places at which it does; there are others at which it points east of north; and others at which it points west of the true north and south line. Its deviation, or in other words what it lacks of pointing in a true north and south line is called its *variation.*

If those places on the earth's surface at which the needle points due north and south be joined by a line, this line is called *the line of no variation.* This line goes quite around the globe in a north and south direction. It is, however, an irregular line, bending now to the eastward and then to the westward. We may trace its general course through North America, by remembering that it strikes the continent near Cape Lookout, on the coast of North Carolina, passes through Staunton, in Virginia, a little east of Cleveland, in Ohio, across Lake Erie, and thence onward to Hudson's Bay.

At places east of this line the variation is toward the west; at places west of it the variation is toward the east.

4. *The variation is subject to annual and daily changes.*—The variation of the needle at any place is

continually changing. For example: the variation at Washington, D. C., was 36′ west in the year 1800, but in 1860 it had increased to 2° 54′. Such a change is going on year by year at all places. The variation increases for several years and then again diminishes. So the needle vibrates, first westward, then eastward, and back again, taking many years to make a single vibration.

Besides this annual variation, the needle has a daily variation, much greater in summer than in winter—amounting to about 15′ in the former and only about 10′ in the latter. At about 8 A. M., the north pole begins to swing westward, and this motion continues until about 1 P. M. Soon after this time it slowly moves back toward the east until, at about 10 P. M., it has reached its starting point. It then moves west again until about 3 A. M., after which it swings back to the eastward until 8 A. M. It completes these two full vibrations every twenty-four hours.

(101.) If a magnetic needle be allowed to move freely up and down, it will seldom rest in a horizontal position. Its inclination is called the dip of the needle.

In the northern hemisphere the north pole of the needle dips; in the southern hemisphere the south pole dips.

1. *The dip of the needle.*—If a slender steel needle be balanced upon a horizontal axis so that it may freely move up and down, and *be then magnetized*, it will be no longer balanced: the north pole will sink until the needle takes a position very much inclined (Fig. 134). The amount of this inclination is called the *dip of the needle.*

In the southern hemisphere the needle also takes an inclined position, but there it is the south pole that points downward.

As the dipping needle is carried farther north, the dip increases until a point is reached where the needle stands in a *vertical* position. Of course this point must be the north *magnetic* pole of the earth: it is a curious fact, that it is not the same as the *geographical* north pole. It is in latitude 70° 5′ N. and longtitude 96° 45′ W.—a little north and west of Hudson's Bay. It was found by Captain Ross, in the year 1832.

Fig. 134.

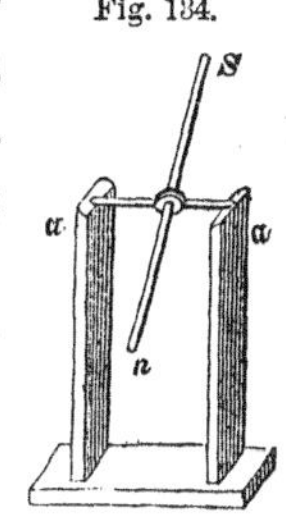

Then, traveling southward in the southern hemisphere, the south pole of the needle dips more and more, and there is evidently a south magnetic pole of the earth. This point has never yet been found. (See Silliman's Phys., Chap. III., § 1).

§ 3. VOLTAIC ELECTRICITY.

(102.) Voltaic electricity may be obtained by means of an apparatus called the voltaic circuit, or better, by means of a Grove's or a Bunsen's battery.

1. *Voltaic electricity.*—We may define voltaic electricity to be electricity which is produced by chemical action. Let us remember that the chemical properties are those which a body may not show without undergoing some change in its nature. So by chemical action, we mean an action by which some change is produced in the nature of bodies. It will not be forgotten [see (3.)] that natural philosophy treats only of the physical properties of matter, and of phenomena in

which there is *no change* in the *nature* of bodies. The student of natural philosophy must not, therefore, expect to find a full explanation of the production and effects of voltaic electricity. Many of its effects, however, do belong to this science.

2. *The voltaic circuit.*—The voltaic circuit, by which this kind of electricity may be produced in the most simple way, is represented in Fig. 135. Into a glass vessel is put a quantity of water, mixed with a little sulphuric acid. A strip of copper and another of zinc are inserted into this liquid, and from the upper ends of these strips two metallic wires project. Now, when the ends of these two wires are brought together, a chemical action will take place between the water and the zinc, by which voltaic electricity will be produced. A multitude of little bubbles of gas rising alongside of the copper strip shows the action of the electric force; and if the ends of the wires be carefully separated, a very delicate spark may be seen between them.

Fig. 135.

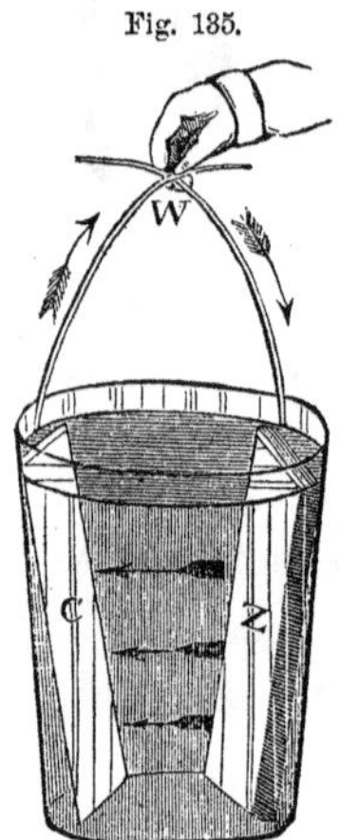

The ends of the two wires are called the *poles* of the circuit: that from the copper strip is the *positive* pole; the one from the zinc strip is the *negative* pole.

3. *Grove's battery.*—In Grove's battery we have a means of obtaining a more powerful action of electricity than in the simple circuit, just described. Two metals, zinc and platinum, and two liquids, dilute sulphuric acid and nitric acid, are used. The peculiar mode of putting these together may be understood by

an attentive study of Fig. 136, aided by the following description:—

A glass vessel (V) is partly filled with dilute sulphuric acid. Into this fluid is placed a zinc cylinder (Z) with a slit from top to bottom, to allow the fluid to circulate, both inside and outside of it, freely. Inside of the zinc cylinder is put a *porous* earthenware cup. Into this cup is poured strong nitric acid, and a strip of platinum is inserted in this fluid. One wire being fastened to the zinc, and another to the platinum, form the poles; and when brought together, and then carefully separated, a spark of electricity may be seen.

Fig. 136.

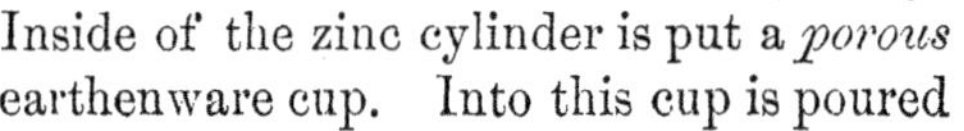

The platinum pole is *positive;* the zinc pole is negative.

4. *Bunsen's battery.*—Bunsen's battery differs from the one just described, by having a carbon cylinder, or rod, in place of the strip of platinum. This does not greatly diminish its action, while it makes it much cheaper, because platinum is a very costly metal.

(103.) In all these forms of apparatus the electricity is produced by chemical action.

A part of its force is overcome by the resistance it meets in going through the poor conductors of the circuit. Its quantity and intensity depend—the first on the size of the metallic plates used, the second on the number of plates employed.

1. *It is produced by chemical action.*—In the voltaic circuit a chemical action takes place between the zinc and the water. The zinc separates the water into two substances—gases, called oxygen and hydrogen. The

hydrogen escapes: we see it as small bubbles which rise along the side of the copper plate; but the oxygen combines with the zinc. By these chemical actions the electric force is produced.

In Grove's and in Bunsen's batteries there is the same decomposition of water by the zinc; in addition to this the nitric acid is decomposed. There is, therefore, much more chemical action than in the simple circuit, and consequently, we should expect a greater electric force produced. For a full explanation of these chemical actions we must refer to the science of chemistry.

2. *The resistance it meets.*—The difference between conductors and non-conductors of electricity, has already been given. We must now attend to the fact that no substance is a *perfect* conductor. Even silver and copper wires, which are among the very best conductors, do not allow electricity to pass through them with perfect freedom; they *resist* its action at every point. So, too, the materials of which the batteries are composed, being imperfect conductors, resist the action of the force at every point, in the fluid, in the plates of metal, and in the wires which join them.

The resistance which a wire offers to the action of electricity through it, depends upon the material of which it is made, and upon its size. The metals are the best conductors, silver standing at the head of the list, copper next, and lead being among the poorest. The larger the wire, the less resistance it offers.

The electric force evolved by the chemical action in the battery must, therefore, exert itself first to overcome the resistance it meets, and the force not thus expended may be used for other purposes.

3. *Its quantity and intensity.*—If large plates of metal are used in the battery, so that the surface, on which the chemical action takes place, is greater, the *quantity* of electricity evolved will be increased, and yet it will be found no better able to overcome resistance than before. If, for example, a single Grove's battery of common size, will give electricity which will go, as a spark, through a layer of air $\frac{1}{32}$ of an inch between the poles, it will be found that another cell, with plates of metal four times as large, will not give a spark any longer than $\frac{1}{32}$ of an inch.

The power of electricity to overcome resistance, is called its *intensity*. It does not depend upon its quantity. There may be a great quantity of electricity with very little intensity, or there may, on the other hand, be a very small quantity with great intensity. We are familiar with something analogous to this in heat, and it may help us in getting a clearer idea of the difference between quantity and intensity. In a burning match there is a small quantity of heat, but it is very intense; while in a large cask of warm water there is a great quantity of heat, with very little intensity.

4. *Quantity depends upon size of plates.*—By increasing the size of metallic surfaces in any circuit, the quantity of electricity is increased. This may be done in two ways: 1st, by having a single plate of each metal made large; or, 2d, by having several small ones of each kind joined together. The last mode is the one usually employed. If in several cells of Grove's battery all the platinum strips are joined, they will be equivalent to one large platinum surface. If all the zinc cylinders be joined together they form one large

surface of zinc. If, then, a wire from the platinum be brought in contact with another from the zinc, the circuit will be completed, and a battery for *quantity* will be formed.

5. *Intensity depends upon the number of plates.*—But if several cells of the battery are joined by connecting the platinum of one cell with the zinc of the next; the platinum of this with the zinc of the next, and so on, finally joining the platinum of the last with the zinc of the first, the *intensity* of the electricity will be vastly increased.

When the resistance to be overcome is considerable, an intensity battery will be used; in most other cases the quantity battery is employed. (See Silliman's Physics, p. 581.)

The greatest difference between voltaic electricity and frictional electricity is this: voltaic electricity is remarkable for its *great quantity* but *feeble intensity*, while frictional electricity is equally remarkable for its *great intensity* but *small quantity*.

(104.) Among the mechanical effects of voltaic electricity, we notice heat, light, magnetism, and induction.

Electricity produces heat whenever it is resisted in its action. It makes the most intense light by passing through air between two charcoal points. It produces magnetism by acting through a wire which encircles a bar of iron. Upon this principle the electric telegraph and other useful instruments are made. By changing the direction of the force around the bar, the poles of the magnet are reversed. One conductor of electricity induces electricity in another near it.

1. *Heat.*—Electricity, when resisted in its action,

shows itself as heat. When it acts through a fine wire, the wire may be made red-hot, and in many cases melted, by the heat produced. Several inches of fine iron wire may be thus melted by a battery of 12 or 15 cells. This power of electricity is applied to the exploding of gunpowder, for blasting rocks. For this purpose, a cartridge is made by filling a tin tube with gunpowder, and corking its ends tightly. Through one of the corks two copper wires pass, joined in the powder by a fine steel wire soldered to their ends. The copper wires are then connected with the poles of a distant battery. The instant that the circuit is made, the fine wire in the gunpowder is melted, and its heat explodes the gunpowder.

No other artificial heat can be made so intense as that produced by electricity. But since it is associated with intense light, we will notice it in that connection.

2. *Light.*—When the wires which lead from the poles of a powerful battery are tipped with charcoal points, if these points are brought in contact and then separated for a short distance, the space between them will be bridged over by an arch of blinding light. On examination, this light is found to be due to the intense whiteness of the carbon tips, chiefly, but *not to their combustion* at all, since, in a vacuum, where combustion can not occur, the light is of equal intensity.

The electric light is unsteady; for this reason, and because of its unpleasant brilliancy, it has not been successfully used for lighting streets and public places. It has been used, however, with excellent effect in light-houses.

The heat of this arch of light is wonderfully intense. Platinum, more difficult to melt than other metals, melts in this heat like wax in the flame of a taper.

Even quartz, and other bodies equally difficult to melt, are fused by it readily.

3. *Magnetism.*—Bars of soft iron inclosed in coils of wire are called *electro-magnets.* The coil is generally called a *helix.* If the two ends of the coil be fastened to the poles of a battery, the electricity darts instantly through the coil, and the iron becomes a magnet. The bar of iron may be of any form; when in the shape of the horseshoe magnet, the coil is made in two parts, one encircling each arm of the iron. A horseshoe electro-magnet A B, is seen in Fig. 137.

Fig. 137.

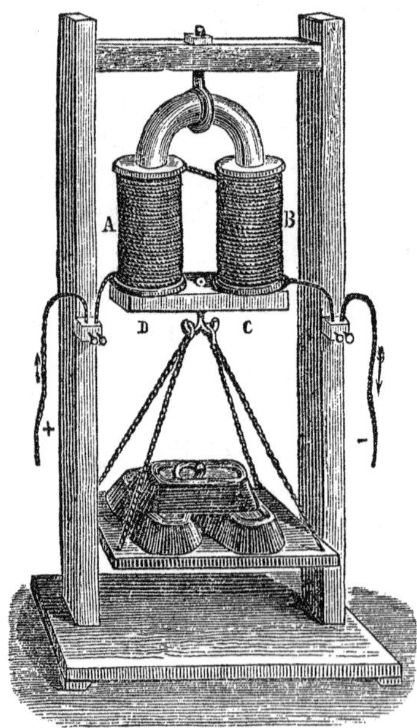

The strength of electro-magnets is something surprising. One belonging to Yale College, weighing 59 lbs, lifted a weight of 2,500 lbs. (See Silliman's Phys., p. 610.) This wonderful power is developed only when a bar of soft iron, the armature, C D (Fig. 137), is in contact with the poles. Without this, the magnet will not lift a tenth part of what it could otherwise sustain.

The iron is magnetic only while the electricity acts around it: let the circuit be in any way broken, and the grasp of the giant is at once loosed: the load falls. On again making the circuit, the magnet is instantly as strong as before. The rapidity with which an iron bar will thus receive and part with its magnetism, as the circuit is made and broken, is truly astonishing. By the electric register for vibrations [see (46.) 3], the author has caused an electro-magnet to undergo this change at the rate of 8,400 times a minute.

4. *The electric telegraph acts on this principle.*—It is upon this principle that the electric telegraph has enabled man to send his thoughts, with lightning-speed, across continents, and under oceans, to his most distant fellow-men.

Having found that a bar of iron will become magnetic as often as electricity is sent round it, and cease to be so on the instant the force stops, let us next notice that the wires conveying the force may be of any length, even miles, and hence the battery may be in one city, while the magnet may be in another, and still an armature will be drawn against its poles every time the circuit is made. Now, if the motion of an armature to and from the poles can be made to *write*, then can messages be sent from one city to another.

The apparatus consists of three parts: the *key*, the *line*, and the *register*.

The *key* is an instrument by which the circuit can be made and broken at will. It is in the office *from* which the message is to be sent. A brass lever, L (Fig. 138), moves on an axis, A. Two projections, *n* and *m*, from its lower side, are just above two others, one of which (*a*) is joined by a wire with the battery, while from

Fig. 138.

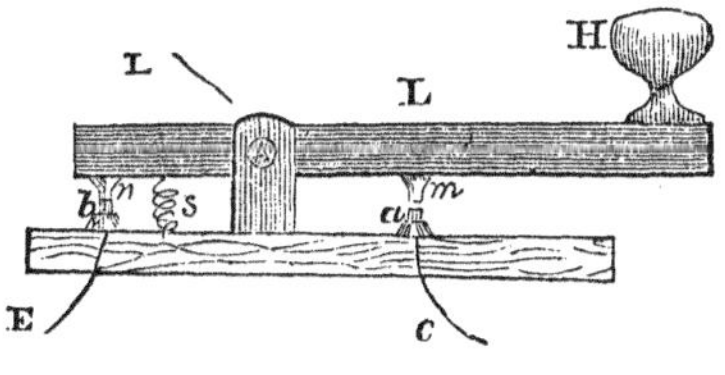

the axis A, another wire reaches to the distant station. By pressing the finger on the end of this lever, the point is brought in contact with the battery wire at *a*, and the electricity can then act through the lever, from the battery wire to the wire from the axis. Let the finger be lifted, and the lever will rise by the action of a spring, *s*, and the circuit is broken.

The *line* consists of a wire (L) reaching from the key over the country to distant places. Formerly two wires were used, one from the positive pole, the other from the negative pole of the battery; but it has been found that the earth may take the place of one of these wires.

The *register* is shown in Fig. 139. One of the screw

Fig. 139.

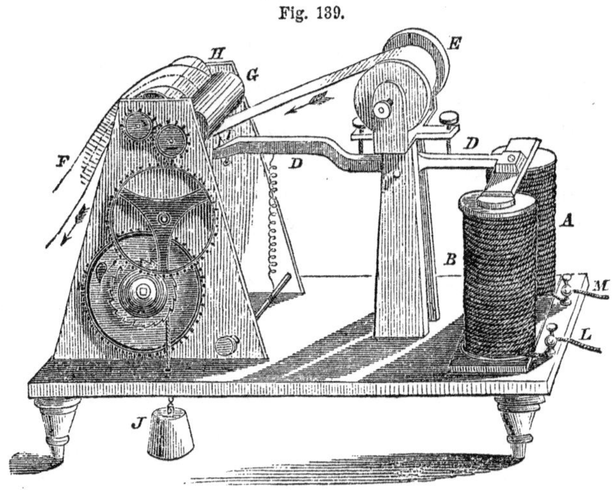

cups at the end of the instrument is connected with the line wire L, from the key of the distant station, while the other M, is connected with the earth. When the circuit is made, the electric force darts around the electro-magnet, and draws the armature down against its poles: this raises the long arm of the lever, and presses the steel point I, against a strip of paper, which is pulled along from the spool E, by clock-work. When the circuit is broken, the armature is released from the poles of the electro-magnet; the long arm of the lever falls by its own weight, or by the force of a spring, and the point is removed from the paper.

If the point press the paper for an instant only, a dot will be made, but if it be held against it for a longer time, a dash will be left upon it. Now, the letters of the alphabet are represented by dots and dashes. Two operators who know this alphabet, can communicate with each other; one by pressing the key causes a series of dots and dashes, to be marked upon the paper of the register at a distant place, while the other can read this written language.

Such is an outline of the *essential* parts of the electric telegraph (see Silliman's Phys., p. 616), the greatest triumph of modern science.

The steam-engine and the electric telegraph may be regarded as the body and the spirit of modern civilization, the first distributing *matter*, the second *thought;* both laboring toward a more general diffusion of comfort and knowledge and sympathy among men.

5. *The poles are reversed by changing the direction of the force.*—Every electro-magnet has a north and a south pole, but they may be made to change from end to end instantly, by changing the direction of the electricity in its passage round the bar. We know that the electricity from Grove's battery always acts from the platinum through the wires to the zinc: it will *enter* a coil by the wire which comes from the platinum; it

will *leave* the coil by that which goes to the zinc. By this means it is easy to trace the direction of the force through the coil. Notice again, that when the ends of the coil-wire are made to change places at the poles of the battery, the direction of the force around the coil is reversed: if it were going around in the same direction that the hands of a clock travel around the dial, it will, on changing the poles, instantly go in the direction which the hand goes when the clock is being set back. If these two points are understood, we may give the following law:—The *south pole* of the electro-magnet will be at that end of the coil where the electricity *enters*, when the force acts around the coil in the same direction which the hands of a clock move over the dial.

6. *One conductor induces electricity in another.*—That a wire through which electricity is acting excites electricity in another one near it, was proved by Faraday in the following way. Two very long, silk covered, copper wires were wound into a coil, so that they should run side by side throughout their entire length, but yet be perfectly insulated from each other. When the two ends of one of these wires were connected with a battery, while the two ends of the other were connected with a galvanometer—an instrument for detecting the presence of voltaic electricity, the galvanometer announced the action of the force in the wire which had no connection whatever with the battery. The direction of this action is the same as in the battery current. When the circuit of the battery was broken, the galvanometer announced a current in the other wire going in the opposite direction.

These currents are called secondary or *induced* currents; they are but momentary, but are renewed at every interruption of the battery circuit.

If a bar of soft iron be thrust into the helix, the force of the induced current is greatly intensified.

It is upon this principle of induction that the *Rhumkorf Coils* are constructed (see Silliman's Phys., p. 627). In these important instruments there is first a coil of large copper wire, inside of which is put a bundle of iron wires. Outside of this, called the primary coil, is placed another, called the secondary coil, made of fine copper wire, from 10,000 to 80,000 feet in length. The two coils are insulated from each other with the utmost care. The ends of the primary coil are attached to the battery, while from the ends of the secondary coil the electricity is taken in experiments.

The effects produced by this apparatus are beyond comparison more intense than from a battery alone, or an electrical machine. When the battery circuit is rapidly made and broken, a torrent of brilliant sparks leap from one end of the secondary coil to the other—in one of the American instruments, a distance of *sixteen inches*. The shocks caused by it are dangerous, in a degree approaching those of the lightning stroke. The heat produced is very intense, while the light obtained by passing its electricity through rarefied air, or through various gases, is beautiful beyond description.

CONCLUSION.

(105.) We have now become acquainted with the important first principles of natural philosophy. Our attention has been given exclusively to those truths which, because of their importance in the theories of science, or because of their practical applications to the wants of life, are essential to be known by those who would be prepared for the higher advantages of the college, and equally necessary to those who expect to become intelligent members of society without the higher courses.

We first learned that the study of natural philosophy, in distinction from chemistry, describes, and attempts to explain, all those phenomena in which there is no change in the nature of bodies.

We then found that the words *molecule*, *inertia*, *attraction*, and *repulsion*, express the four fundamental thoughts which open the way to an explanation of all the phenomena of which the science treats.

In the application of these thoughts we noticed *first*, that attraction and repulsion, acting on the molecules of matter, produce the three physical forms, solid, liquid and gaseous, and we were then able to describe the characteristic properties of these groups of bodies.

Second.—That attraction and repulsion, acting upon masses, cause the phenomena of rest and motion: and this introduced us to the explanation of the laws which control the wonderful variety of motions in nature and the important action of machines.

Third.—That attraction, repulsion, and inertia, acting upon the masses or molecules produce vibrations, and we were able to obtain laws for this peculiar kind of motion. And when we noticed, further, that the vibrations of molecules affect our senses, we had found the key to unlock the gates which opened into the fields of *sound*, of *light*, and of *heat*, and were then able to explain their wonderful effects.

Fourth.—That the constantly opposite action, or struggle, of attraction and repulsion among the molecules of bodies shows itself as electricity, and, following this thought, we became acquainted with the curious effects of electrical action.

Nor are the applications of the four fundamental thoughts limited to the phenomena which have been presented in the compact outline of the science just completed. The most extended commentary—one which should give *all* the explanations of those phenomena in which there is no change in the nature of bodies—if written, would only *fill up* the scheme which these four thoughts suggest. Indeed, they seem to be the frame-work upon which the fabric of material things has been built.

One who has never given special attention to the study of nature, finds his mind overwhelmed by the great diversity of material objects which the world presents, and he beholds them with amazement, or it may be with indifference; but, in the light of careful observation and analysis, the system of material things is simple and orderly, displaying the infinite knowledge, power, and skill of a Divine Architect.

APPENDIX.

—o—

I.—SOUND.

(CHAPTER V. CONTINUED).

§. 4. MUSICAL AND SENSITIVE FLAMES.

(106.) WHEN a gas flame burns within a glass tube, a musical sound is produced. The pitch of the tone depends on the length of the tube and the size of the flame. A silent flame may be made to sing by sounding near it, the note of the tube. Naked flames are also sensitive to the action of neighboring sounds.

1. *The musical flame.*—Let a flame of common coal gas, placed under the end of a glass tube T (Fig. 140), be slowly raised into it; when a particular height is reached, the flame, if small enough, will burst forth into a loud and continuous sound. This sound is often harsh, sometimes melodious. At the beginning it is sometimes low and smooth, like the whistle of a very distant locomotive, but as the experiment goes on, the intensity of the sound rapidly increases until, like the long monotonous screech of the engine at hand, it becomes almost unbearable.

In tubes of tin and pasteboard, sounds of different quality are obtained.

That a gas flame flutters when exposed to a gentle

breeze, is a fact sufficiently familiar. Now, this fluttering flame, like a vibrating tongue or reed, must cause vibrations in the air around it. Here is the key to the explanation of musical flames.

Fig. 140.

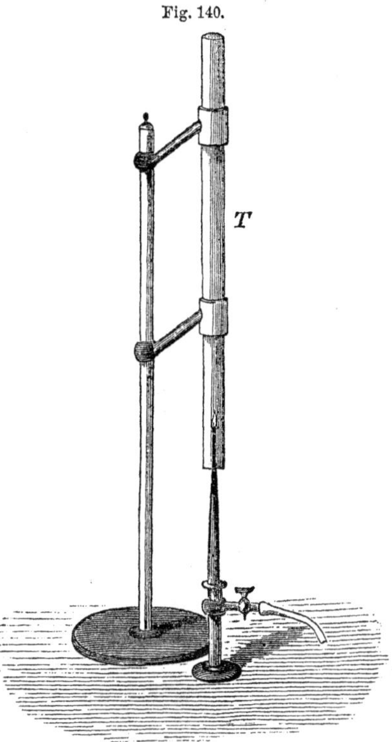

The air in the tube is heated by the flame; it rises: an upward current through the tube is thus produced; the flame flutters in this current, and causes a system of waves, whose rapidity and amplitude give pitch and intensity to the note produced. If we inquire further about the cause of the fluttering, we are told that experiments by Faraday proved that gas issues from a burner in an *unsteady* stream, due to the friction against the sides of the tube, and that in burning, it makes a series of inaudible explosions. A current of air heightens both of these effects, and makes them sensible.

That a musical flame is thus intermittent is shown by the following beautiful experiment. The tube T (Fig. 141), is blackened so as to keep the light from falling on the screen placed behind it at S. A concave mirror M, in front of the flame, forms an inverted image of it on the screen. If the mirror is turned horizontally,

Fig. 141.

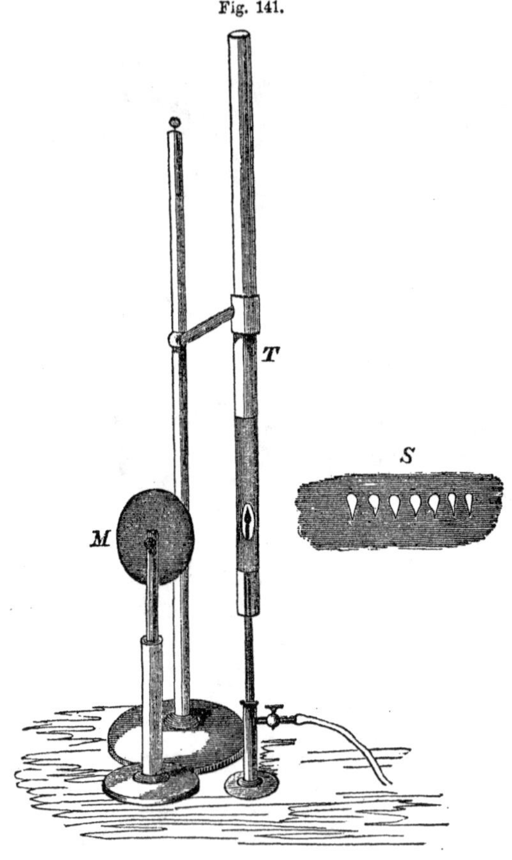

while the flame is silent and steady, the image will move, and if the motion of the mirror is swift, an unbroken band of light will be seen on the screen. But, if when the flame is singing, the mirror is swiftly turned, a series of *distinct images* will appear.

This experiment teaches, that in the act of singing, the light of the flame is quenched at intervals. And if, as Dr. Tyndall supposes, the spaces between the images are absolutely dark, then the flame must be entirely put out at intervals, the heat being sufficient to instantly relight it.

2. *The pitch of the note.*—In these tubes, as in organ pipes, the pitch of the sound depends upon the length of the tube. But while the pitch depends chiefly upon the length of the tube, it is partly governed by the size of the flame. If one tube is just twice the length of another, its fundamental note is an octave below, but when placed over a flame whose size fits it to sing in the shorter tube, the note of the shorter tube will be produced. Then let the flame be gradually enlarged and in a little time, the low fundamental note of the tube suddenly bursts forth. By varying the size of the flame it is possible to obtain the fundamental note, its octave, and its four harmonics from the same tube.

3. *A silent flame responds to a sound.*—At one place in the tube a flame spontaneously bursts into sound; at the other, near the end, it trembles, but does not sing. Now, between these, there is a third place, at which the flame is silent, but where, if the proper note is sounded, "it stretches forth its little tongue and begins its song." Dr. Tyndall goes on: I stop the music; and now standing as far from the flame as the room will allow me, I command the flame to sing. It obeys immediately. A pitch, pipe or any other instrument which yields a note of proper pitch, produces the same effect.

4. *Sensitive flames.*—The name "sensitive flames" is

given to those which, without being inclosed in tubes are affected by sounds. Certain sounds in an instrumental concert cause curious motions, often, of the gas flames in the room. This observation was first published in 1858. The motions referred to consist of a "jumping" of the flame to considerable height, or a thrusting forth of tongues of flame from its upper edge. (See Tyndall, On Sound p. 230.)

An effect just opposite this, the shortening of tall flames, was first noticed by Mr. Barrett of London, in 1865. He says (Chem. News, Amer. Rep., July, 1868):—"A jet of gas issuing from a V-shaped orifice was found to be quite insensible to sound until the flame reached a height of 10 or 12 inches (see Fig. 142), and then, at the sound of certain high notes, the flame shortened and spread out into a fan shape." Another flame he thus describes:—"So sensitive is this flame that even a chirp made at the far end of the room brings it down more than a foot. Like a living being, the flame trembles and cowers down at a hiss—it crouches and shivers as if in agony at the crisping of this metal foil, though the sound is so faint as scarcely to be heard. It dances in tune to the waltz played by this musical box—and, finally, it beats time to the ticking of my watch."

Mr. Barrett also suggests that these flames may yet be turned to some use, and to illustrate, suggests an arrangement shown in Fig. 142.

Near the tall sensitive flame *a*, stands two vertical brass rods, *b c*. Projecting from these rods are two metallic ribbons, made of layers of silver, gold, and platinum welded together. The ends of the ribbons are about half an inch apart. By heat the different metals expand unequally, and, bending the ribbons, bring their ends together. The brass rods are connected by wires with an electric bell at a distance.

Fig. 142.

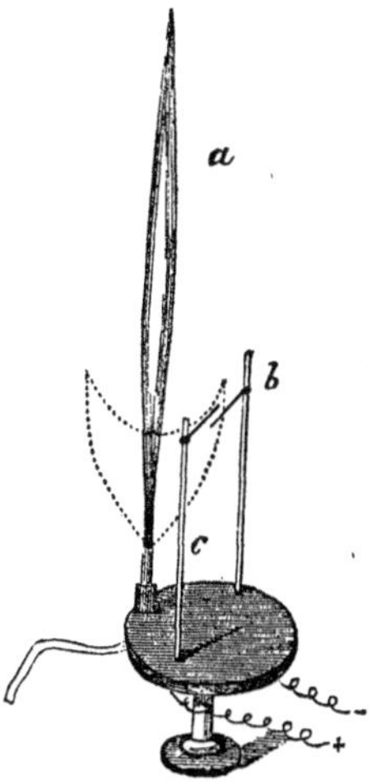

Now, as long as the tall flame is not disturbed, the metallic ribbons are not in contact: the circuit is broken, and the bell is silent; but at the sound of a whistle the flame jumps down, warms the ribbons, completes the electric circuit, and rings the distant bell. A flame, sensitive to the sound of burglars' tools, might in this way sound an alarm.

Another quotation from the same lecturer may end this subject. "Imagine that when enchanted by the performance of some well executed opera or oratorio, a companion by our side were to say—'Well, after all, of what good are these fine sounds? to what practical end can you turn this music?' Should we not instantly condemn a speech so characteristic of a sordid and sensuous mind? And when the student of nature is listening with admiration, and even awe, to the sweet though silent music sung to him by every object of his diligent study—by air and water, by flowers and flames, he is conscious that he bows before an oratorio as far above that of Handel, as the works of the Creator are superior to the compositions of the creature."

§ 5. METHODS OF REGISTERING VIBRATIONS.

(107.) Various methods have been devised for registering vibrations. The syren shows the number of air puffs corresponding to the sound it makes. Savart's wheel tells the number of taps of a solid body to produce a continuous sound of any pitch. In Duhamel's graphic method, the vibrating solid traces a sinuous line: in Scott's phonautograph this method is applied to all *sonorous* vibrations. By the author's electric register, vibrations whose amplitude are appreciable, either in solids or liquids, may be directly registered.

1. *Registering vibrations.*—Sound, light, heat, and perhaps electricity, are caused by vibrations. The study of these subjects, then, whenever we pass beyond the mere description of sensible phenomena, is the study of the nature and laws of vibrations. The nature of the vibrations which produce heat and light, can only be inferred from the phenomena they cause. Almost inconceivably rapid, they can be neither counted nor measured by any direct experiment. What agent swifter than they, able to magnify their amplitude and mark their number! Electricity is swifter than light: will the future develop some means by which it may keep pace with, and mark the ebb and flow of luminous waves? The boldest experimenter would yet hardly dare venture the conjecture.

Sound waves have, however, been registered by direct experiment.

2. *The syren.*—This instrument, already described [See (56.)], registers the number of air puffs by which its sound of any pitch is made, and the number of puffs is taken as the number of vibrations. Improved forms of this beautiful and accurate instrument are described in Tyndall's Lectures "On Sound."

3. *Savart's wheel.*—In this instrument a toothed

wheel is turned by another wheel and band. A thin metallic tongue is so placed that the end of it presses gently against the teeth of the wheel. When the wheel turns, this tongue taps against every tooth, and when turned fast enough, these taps are linked together into a continuous sound. An index, attached to the axis of the toothed wheel, shows the number of taps, and the number of taps corresponding to any sound is taken as the number of vibrations to produce it.

On the principle that sounds of the same pitch are made by vibrations of the same rapidity, both the syren and Savart's wheel may *indirectly* register the number of vibrations made by any sounding body. Pitch either of these instruments in unison with the sound of the piano, for example, and its dial tells the number of vibrations made by the wire. Owing to the difficulty of judging an exact unison, and to the greater difficulty of keeping a perfectly uniform motion of the instruments, a direct registry would seem to be more satisfactory.

4. *The graphic method.*—The graphic method of Duhamel consists in fixing a fine metallic point to the body emitting the sound, and causing it to trace the vibrations on a properly prepared surface.

The apparatus consists of a cylinder, around which is rolled a sheet of paper covered with a thin film of lamp-black. Suppose the vibrations of a tuning fork are to be registered. The handle of the fork is set in a vise, or other firm support: a fine point is cemented to one of its prongs, and the cylinder is then so placed that the point will gently rest upon its surface. Now, if the cylinder is turned, the point will trace a straight line, by scraping off the black film in its path over the white paper; but when the fork vibrates, the point moves with it, and instead of a straight line, it traces a sinuous path, each undulation representing a double vibration. By counting the number of undulations made in given time, the number of vibrations is known. (Atkinson's Ganot's Phys., p. 159.)

5. *The phonautograph.*—M. Leon Scott has applied his graphic method to all *sonorous* vibrations whatever. His apparatus, called the phonautograph, consists of a box about a foot and a half long, and one foot in its greatest diameter, having the shape of a barrel or cask. It is made of plaster of Paris, one end being open, the other closed by a solid head, to the middle of which is fitted a copper tube having a thin membrane stretched over its outer end. Near the center of this membrane is fixed a very light projecting point, in front of which turns the cylinder covered with blackened paper, the same as in Duhamel's method.

Now, whenever a sound is made in front of the cask, the air in it, the membrane, and its projecting point, will vibrate in unison with it, and the number of undulations in the path traced upon the cylinder is the number of vibrations made. By this ingenious method the number of vibrations made by the voice in singing, or indeed by any noise whatever, if of sufficient intensity, may be directly registered. (Atkinson's Ganot's Physics, p. 190.)

6. *The electric register.*—The electric register [See (46.)] may give a direct and an almost unerring registry of the vibrations in solid or liquid bodies, where the amplitude of vibration is at all appreciable, whether accompanied by sound or not.

According to the experiments of Wheatstone the passage of the electric spark in the discharge of the Leyden-jar occupies $\frac{1}{24000}$ of a second. If passed through paper moistened with iodide of potassium, this spark has time to decompose this substance, and leave a brown stain upon the paper. When a little starch is mixed with the iodide, the stain is blue and very distinct. The possibility of registering at least 24,000 *double* vibrations a second is thus distinctly pointed out. The requirements are: 1st, a steady stream of electricity, from a powerful battery; 2d, means by which the motion of the vibrating body may open and close this circuit; and 3d, a *rapid* motion of the chemically, prepared paper, through which the electricity is passing.

With an apparatus of somewhat rude construction, applied directly to the wires of a piano in daily use, the following results have been obtained.

Note........	C,	C♯,	D,	E♭,	E,	F,	F♯,	G,	A♭,	A,	B♭,	B,	C,
Vib. per Sec.	64.2,	67.1,	73.1,	76.1,	81.8,	87.1,	92.8,	98.1,	101.5,	107.8,	114.5,	117.5,	127.5.

These numbers are in each case the mean of several experiments, but in no set of experiments did the registry vary except by *a single mark*, which must correspond to an error of less than one-half a vibration.

The C, 127.5, is written in the second space of the base.

II.—LIGHT.

(CHAPTER VI. CONTINUED).

§ 6. INTERFERENCE OF LIGHT AND WAVE LENGTHS.

(108.) Luminous vibrations, by crossing, interfere, and produce waves of different amplitude. The intensity of light depends upon the amplitude of vibration; its color upon the rapidity of vibration. The length of waves of light vary from .0000167 of an inch to .0000266 of an inch. Diffraction fringes are the effect of interference.

1. *The interference of light.*—We have learned that light is the result of vibrations in a very elastic medium called ether. We ought, therefore, to expect that the phenomena of light would, in many respects, be like those of sound and heat. We have found this to be true, their laws of reflection, refraction, and transmission being alike. They are alike also in regard to interference. (See Silliman's Physics, pp. 258 and 273.)

If two sound waves in air cross each other, so as to bring their condensed parts or phases together, they cause a wave whose amplitude is equal to the sum of theirs, and produce a sound of greater intensity. If they come together in such way that the condensed phase of one strikes the rarefied phase of the other, they cause a wave whose amplitude is equal to the difference of theirs, and produce a sound of less intensity.

So, too, if waves of light, in the ether, cross each other so as to bring their like phases together, they cause a wave whose amplitude is equal to the sum of theirs, and produce a light of greater brightness; but if their opposite phases are thrown together, they form a single wave whose amplitude is equal to their difference, and cause a light of less intensity.

Now, examine the conditions of interference more carefully. A wave consists of two phases, and the sum of their lengths is the length of the wave. Of course, then, each phase is just one whole wave length ahead of the next one behind it of the same name, and just one-half a wave length ahead of the next one behind it of a different name. If, then, two sets of waves are to interfere with like phases together, their starting-points must be *one* wave length, or some *whole number* of wave lengths apart; to bring different phases together, the distance between their starting-points must be one-half a wave length, or some multiple of this.

2. *Color depends upon rapidity of vibration.*—As the pitch of sounds depend upon the rapidity of the vibrations which cause them, so the color of light varies with the rapidity of luminous vibrations. A red light is made by the slowest, a violet light by the swiftest, vibrations.

3. *The length of light waves.*—Light of all colors travels with the same velocity; and since violet is produced by the most rapid vibrations, the *length* of its waves must be less than for any other. Suppose, now, that two sets of waves start from surfaces very near to each other, but *not parallel:* at some points the dis-

tance between them will correspond to the wave length for violet; at others, to the wave lengths of other colors. The result of the interference of the two sets of waves will be to form, at different points, all the tints of the spectrum. The rainbow colors of the soap-bubble, which so delighted us in childhood, illustrate this most beautifully. The light is reflected from both the outside and the inside surfaces of the thin film; these surfaces are not parallel; and the interference of the two sets of waves gives rise to the colors.

Now, could we but measure the thickness of the film at the point where red is seen, we would find the length of the wave for red; and if at points where other colors appear, we would find the wave lengths which produce them. Newton actually calculated these minute spaces, although, of course, so frail a thing as a soap-bubble could not be used for the purpose. His plan may be understood from Fig. 143. A very thin layer of air is included between two very smooth glass surfaces, one curved, the other plane. When the glasses are pressed together, a series of rainbow-colored rings are seen, with a black center at the point *a*, where the glasses are in contact. If red light alone is used, a series of red rings will be separated by dark spaces.

Fig. 143.

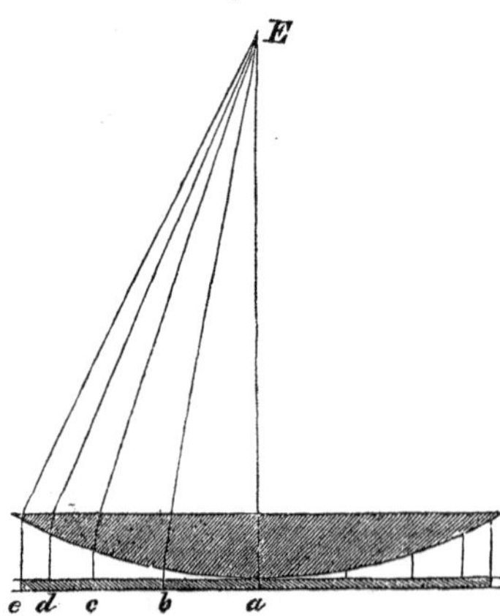

Now these rings are caused by the interference of

two sets of waves, one reflected from the lower side of the curved glass, the other from the upper side of the plane glass, meeting at the eye, E. Newton calculated the thickness of the layer of air at points, *b c d e*, where the rings were seen, and from these thicknesses calculated the length of the waves. The more refrangible colors are produced by shorter waves. The lengths of luminous waves vary between .0000167 of an inch for violet, to .0000266 of an inch for red. (See Silliman's Phys., p. 376.)

4. *Diffraction.*—Diffraction is the change which light undergoes when it passes the edge of an opaque obstacle. Place two knife-blades edge to edge, and look through the narrow slit between them at the clear, bright sky. Instead of a well-defined, clear, bright space, a great number of very delicate parallel black lines will be seen. The edge of a single blade, or of any thin body, will appear fringed with dark lines, and, under some circumstances, with colored bands of great beauty. One who has been taught to recognize it, will be surprised to find how numerous and common are the various forms of this delicate phenomenon. A lady, on suddenly lifting her eyes to the bright sky, sees it, through the meshes of her veil, covered with a net-work of rainbows. Who has not wondered at the brilliant colors of the sky, seen through the fine fibers of a bird's feather?

By the following experiment diffraction fringes may be shown inthe class-room. A beam of sunlight L (Fig. 144), coming through a narrow slit in the shutter of a darkened room, is made to pass through a second slit O, at a distance of ten or fifteen feet. When a white screen, S (a front view shown at S), is placed behind this slit, at a distance of about four feet, a multitude of colored bands, alternating with dark spaces, will appear upon it.

Fig. 144.

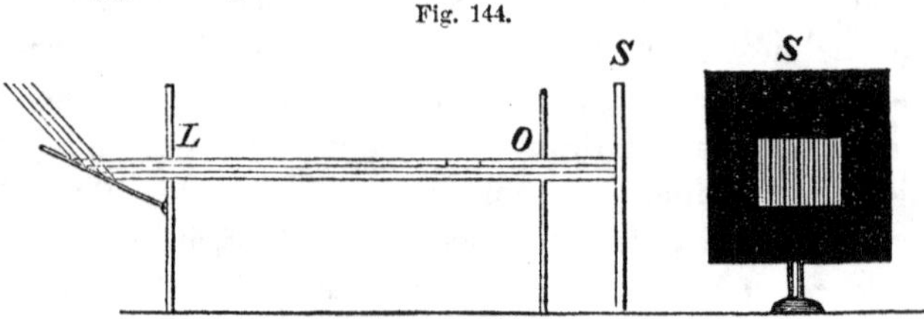

Diffraction fringes are caused by the interference of light. When one set of waves pass an obstacle, it starts another set in the ether on the other side. These two sets going almost, but not exactly, in the same direction, interfere and give rise to the many curious results known as diffraction.

§ 7. DOUBLE REFRACTION AND POLARIZATION.

(109.) When a beam of light passes through a crystal of Iceland spar it is doubly refracted. The two beams which emerge are both polarized. Light may be also polarized by reflection. The effects of polarized light are numerous and important.

Fig. 145.

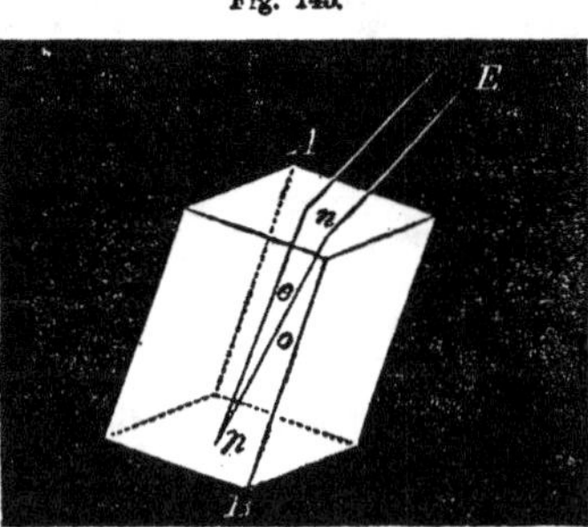

1. *Double refraction.*—Crystals of Iceland spar are found very transparent, and of a form (Fig. 145) as regular as could be cut by the hand of a skillful artist. Each of its six surfaces is a parallelogram. They are so arranged that three

of them have each an obtuse angle at A, and the other three each an obtuse angle at B. A line joining the points A and B is called the *optic axis* of the crystal. Now, if a ray of light be passed through such a crystal in any direction *not perpendicular to the axis*, it will emerge as two separate rays, and the light will be said to be *doubly refracted.*

Thus, suppose a beam of light coming up from below enters the crystal at the point *p* (Fig. 145), it will be divided into two parts, *o* and *e*, which, emerging at points *r* and *s*, go on as parallel and separate beams, and cause the curious effect of making any thing on which the crystal rests appear to be double. One of these refracted beams (*o*) obeys the regular law of refraction; the other (*e*) does not. The first is called the *ordinary* beam, the other the *extraordinary* beam. Many other transparent crystals have this power of double refraction.

2. *Both beams are polarized.*—A very curious change is wrought in the light by double refraction. Common light will pass through any transparent medium, no matter in what position it may be held, but these doubly refracted rays are able to pass through a second medium when it is held in certain positions only. For example, if the ordinary ray be made to fall upon a flat plate of tourmaline (a transparent mineral crystal), and it go through when in one position, it will not go through when the plate has been turned 90° around. Turn the plate 90° more, and the ray will again pass through it; turn it 90° further yet, and the ray will be again wholly cut off.

If the extraordinary ray be tried, it will be wholly transmitted by the plate in positions where the ordinary ray was cut off, and wholly cut off where the other was transmitted.

When light, by being refracted or reflected, is made incapable of being again refracted or reflected except in certain directions, it is said to be *polarized.*

3. *Polarization by reflection.*—If a beam of light, shown by *a b* (Fig. 146), falls upon a plate of glass at an angle of incidence 56½°, a part of it will pass into the glass, the rest of it will be reflected. If the reflected part be examined by a plate of tourmaline it will be found to be polarized. Or if another plate of glass N, is placed parallel to the first, the beam will be reflected as the figure shows it, but let the plate be turned 90°, as showed by the dotted lines, and the beam will be wholly cut off. Turn it 90° farther, and the reflected beam appears again; another 90°, and it is again cut off. At any other angle of incidence than 56½°, the light will be only partly polarized: 56½° is the *polarizing* angle for glass.

Fig. 146.

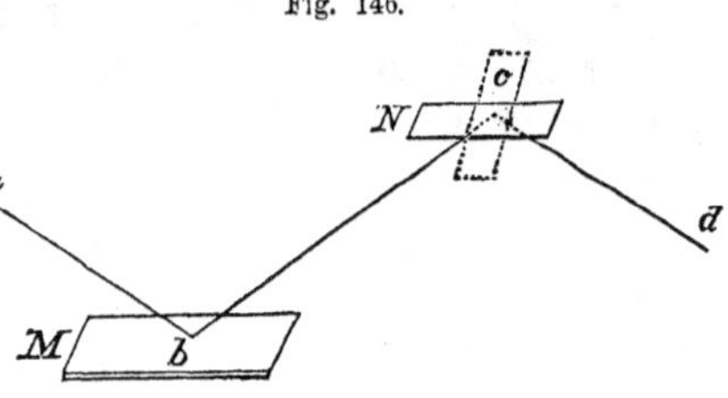

4. *Polarizing instruments.*—The instruments, called polariscopes, by which to study polarized light, consists essentially of two parts, one to polarize the light, the other to examine it after it has been polarized. The first is called the *polarizer*, the second, the *analyzer.* One of the simplest forms of the instrument is shown in the figure (Fig. 147). The polarizer P, is a plate of glass, covered on the back of it with black varnish.

The analyzer A, is a plate of tourmaline set into a movable tube. Objects to be examined by polarized light are supported in a movable ring O.

Fig. 147.

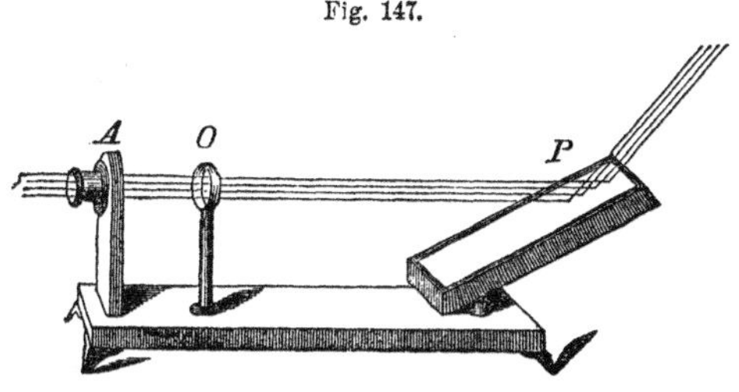

5. *Theory of polarization.*—To explain the phenomena of polarization, we must remember that light is caused by vibrations, and add to this the assumption that these vibrations take place in all possible directions *at right angles* to the *direction in which the ray itself is going.* Let us for a moment suppose that we could see a single ray of light, and that we look squarely at the end of it. We may fancy that we would see a circular outline, with the particles of ether moving swiftly in the directions of all its diameters. Let (Fig. 148) A represent this view. Now the theory assumes

Fig. 148.

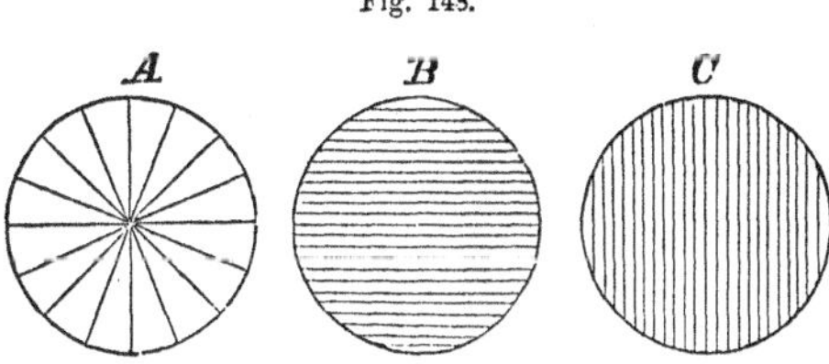

that by refraction or reflection, all these vibrations are changed into two sets, one which vibrates in a horizontal plane (Fig. 148), B, and the other in a vertical plane (Fig. 148), C. This change is what is called polarization. The tourmaline plate, or the plate of glass, will let one of these sets of vibrations pass through it only when in certain positions, owing to some peculiar arrangement of its molecules, but when in position to cut off one set it allows the other to pass freely.

6. *Effects of polarization.*—When a thin plate of mica, or other doubly refracting medium, is put at O, in the polariscope (Fig. 147), the two beams emerging, by interfering, produce most beautiful colors. When seen in certain directions, colored rings of surprising beauty, with a black cross, appear (see Atkinson's Ganot's Phys., p. 510). The form and arrangement of these rings differ in different crystals—a fact of much interest to the mineralogist.

Some substances have the power to change the position of the plane of vibration, in a ray of polarized light. Thus if the analyzer (Fig. 147), is turned so that the ray of polarized light is turned off, a thin plate of quartz at O, will cause the ray to reappear. In this case suppose the vibrations to be in the vertical plane, and that the analyzer is turned to the right just 10°; the quartz must bend the plane of vibration 10° to the right also, in order that the ray may pass. This is called *rotary* polarization.

A great number of liquids have this power. Some of them turn the plane to the right; such is a solution of cane sugar; others turn the plane to the left; such is a solution of grape sugar. This fact is of great interest to the chemist, and it assists the physician at times to determine the healthy or diseased condition of the fluids of the human system.

III.—HEAT.

(CHAPTER VII. CONTINUED).

§ 5. THE MECHANICAL EQUIVALENT OF HEAT.

(110.) As MECHANICAL action produces heat, so in disappearing, heat produces mechanical action. The exchange takes place in definite quantities. The force exerted by a weight of 772 lbs., falling through a distance of 1 ft., produces heat enough to raise the temperature of 1 lb. of water 1° F.

1. *Heat by mechanical action.*—We have seen that friction, percussion, and pressure, are sources of heat. Let us now examine these sources more fully.

When two pieces of wood are rubbed together, we notice that there is, 1st, the *force* of the hand, 2d, the *motion* of the pieces, and 3d, the heat evolved. So, too, when one body strikes another, as when a heavy weight falls upon an iron plate, we may notice, 1st, the *force* of gravitation, 2d, the *motion* of the weight, and 3d, the *heat* produced by the blow. Should we examine other cases, we should find in them all, as in these, that mechanical *force* produces *motion*, and that the motion, when interrupted, evolves the *heat*.

2. *Mechanical action by heat.*—But we may reverse the order. Heat, by producing motion, may exert mechanical force. The steam-engine affords a familiar and sufficient example. In this machine, the *heat* of

the furnace, by the steam it forms, gives *motion* to the piston by which the tremendous *force* of the engine is exerted. Motion is the medium in which mechanical force and heat are interchanged.

3. *In the exchange no loss occurs.*—The motion of a sledge hammer can give rise to a certain amount of motion among the molecules of the anvil on which it falls, no more, no less. This molecular motion appears as heat. Now, if this heat could be all collected and changed back into mechanical force, it would be just sufficient to lift the hammer to the height from which it fell. The amount of heat produced by a given force will always be the same, and when it disappears it can exert a force just equal to that which caused it. Nature permits no loss in her exchanges.

4. *The mechanical equivalent of heat.*—Is it then possible to tell just how much heat will be produced by a given amount of mechanical force, or how much force a given amount of heat may exert? This has been done with the greatest accuracy.

The first step in the investigation was, to settle upon some unit by which to measure the force exerted, and another by which to measure the heat produced. The unit of force chosen, is the force exerted by a weight of 1 lb., falling a distance of 1 ft. The unit of heat, is the amount of heat required to raise the temperature of 1 lb. of water 1° F.

Now the question is, how many units of force will produce one unit of heat? The honor of first answering this question is shared by Dr. Mayer of Germany, and Mr. Joule of England, who, at about the same time, and by different methods, obtained results so much alike as to give impartial judges great confidence in, not less than admiration of, the labors of both. The experiments of Joule extended through seven laborious years. Dr. Tyndall thus speaks of them in his *Heat as a Mode of Motion*:—

"He placed water in a suitable vessel, and agitated that water by paddles driven by forces which he could measure, and determined both the amount of heat, by the stirring of the liquid, and the amount of labor expended in the process. He did the same with mercury, and with sperm oil. He also caused disks of cast-iron to rub against each other, and measured the heat produced by their friction, and the force expended in overcoming it. He also urged water through capillary tubes, and determined the amount of heat generated by the friction of the liquid against the sides of the tubes. And the results of his experiments leave no shadow of a doubt upon the mind that, under all circumstances the quantity of heat generated by the same amount of force is fixed and invariable." The same author goes on to say: "It was found that the quantity of heat which would raise one pound of water one degree Fahr. in temperature, is exactly equal to what would be generated if a pound weight, after having fallen through a height of 772 ft., has its moving force destroyed by collision with the earth. Conversely, the amount of heat necessary to raise a pound of water one degree in temperature, would, if all applied mechanically, be competent to raise a pound weight 772 feet high, or it would raise 772 *pounds one foot high*. The term 'foot-pound' has been introduced to express the lifting of one pound to the height of a foot." Then 772 foot-pounds is what is called the *mechanical equivalent of heat*.

IV.—ELECTRICITY.

(CHAPTER VIII. CONTINUED).

§ 4. MAGNETO-ELECTRICITY.

(111.) Electricity may be developed by a magnet: it is then called magneto-electricity, and the apparatus which develops it is called a magneto-electric machine. All the effects of voltaic electricity may be caused by electricity from a magnet: from Wilde's machine these effects are most extraordinary.

1. *Magneto-electricity.*—We have seen that when a bar of soft iron is inclosed in a coil of wire, a current of electricity renders it magnetic; now just reverse the conditions: put a magnet into a coil, and a current of electricity will be induced in the wire. It flows for an instant only, but is renewed when the magnet is withdrawn. Or, if a bar of soft iron be rapidly made to receive and part with magnetism, a rapid series of electric currents will act through the wire which is wrapped around it.

2. *Magneto-electric machine.*—A great many forms of apparatus for getting magneto-electricity have been devised: one of the most common is shown in Fig. 149. In this instrument two coils of wire (W) inclose the two arms of a bar of soft iron, having the form of a horseshoe magnet, and which by a band and wheel M, can be put in rapid motion in front of a very power-

ful compound magnet, S. The soft iron becomes magnetic whenever its ends are in front of the poles of the permanent magnet, and, hence, its two branches are

Fig. 149.

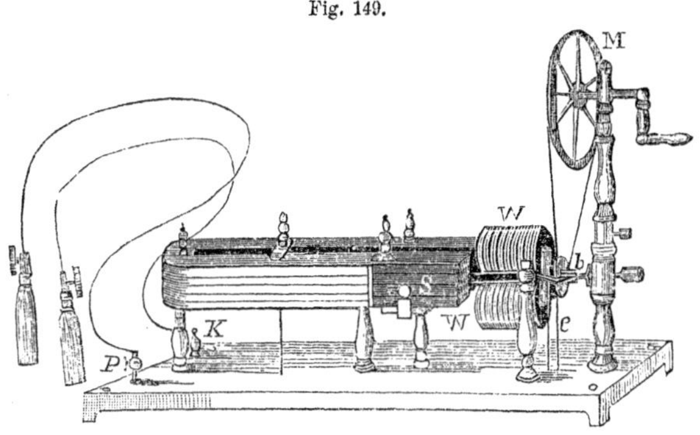

being alternately magnetized in opposite states at every turn. The effect of this is to produce two opposite currents in the coils at every revolution. These currents are taken by the wires *c*, and thence under the instrument to the screw cups, K and P.

3. *Effects of magneto-electricity.*—A person grasping the handles of the machine (Fig. 149), receives shocks, which become unbearable when the motion of the armature is rapid. Chemical decompositions, heat, light, and indeed all the effects of electricity from a voltaic battery, may be obtained by magneto-electricity.

4. *Wilde's machine.*—The electricity obtained from the machine (Fig. 149), may be used to magnetize an electro-magnet, and an armature revolving in front of its poles will give a powerful current, by which a second and still stronger electro-magnet may be mag-

netized. Upon this principle, Mr. Wilde, of England, has invented a machine of great power. Starting with several strong steel magnets in a row, electricity is is taken from the wire of a long armature revolving between their poles, and passed through the coil of a very large electro-magnet. From another armature, revolving between the poles of this magnet, a second current of electricity, some hundreds of times stronger than the first, is taken, and used for any purpose for which the machine is intended. The armatures are turned by a steam-engine.

The most extraordinary effects are produced by this machine. With a small one, having only six small steel magnets, an iron rod, fifteen inches long, and a quarter of an inch in diameter, was melted, and in the light produced by a larger machine, the flames of street lamps half a mile away, were said to have *cast shadows* upon a screen.

§ 5. THERMO-ELECTRICITY.

(112.) Electricity may be developed by heat: it is then called thermo electricity.

1. *Thermo-electricity.*—We have seen that electricity will produce heat; we are now to notice that heat will produce electricity. When two pieces of different metals are soldered together, and their junction heated or cooled, a current of electricity is produced. The metals antimony and bismuth are best suited to this purpose, but any others, or indeed, two pieces of the same metal, will, in some degree, produce the same effect. Nor is it quite necessary that metals should be

used at all; other solids, and even fluids, give rise to this kind of electricity.

Two pieces of metal soldered together, with wires attached to their other ends, through which the electricity may act, is called a *thermo-electric pair.* When stronger currents are desired, a combination of pairs, the two metals alternating throughout, is used. Such a combination is called a *thermo-electric pile.*

Metals which differ most in conducting power and crystalline texture, are best suited to produce thermo-electric currents. The force of the electricity is in proportion to the difference of temperature at the two ends of the pile, provided the difference does not exceed 80° or 90° F. It is in all cases very feeble; yet the galvanometer (an astatic needle inclosed in a helix) responds to so delicate a force, that the slightest change of temperature in the pile can be detected by the current it produces. Indeed the thermo-electric pile and galvanometer (Melloni's apparatus), is the most delicate thermometer known. (See Silliman's Phys., pp. 636 and 443.)

THE END.

www.ingramcontent.com/pod-product-compliance
Lightning Source LLC
LaVergne TN
LVHW020231110826
845151LV00003B/892

* 9 7 8 1 4 2 5 5 3 1 4 1 6 *